SUSTAINABLE DRYING TECHNOLOGIES

Edited by **Jorge del Real Olvera**

Sustainable Drying Technologies

http://dx.doi.org/10.5772/61766
Edited by Jorge del Real Olvera

Contributors

Saban Pusat, Mustafa Tahir Akkoyunlu, Hasan Hüseyin Erdem, Liliana Alamilla-Beltrán, Jaime Jiménez-Guzmán, Diana E. Leyva-Daniel, Brenda H. Camacho-Díaz, Antonio R. Jiménez-Aparicio, Ester Betoret, Laura Calabuig-Jiménez, Cristina Barrera, Marco Dalla Rosa, Tomasz Pniewski, Marcin Czyż, Jorge Del Real Olvera

CBS, Edition **2017**

Published by InTech

Janeza Trdine 9, 51000 Rijeka, Croatia

Publishing Process Manager Dajana Pemac

Technical Editor SPi Global

Cover Designer InTech Design Team

Additional hard copies can be obtained from orders@intechopen.com

Sustainable Drying Technologies, Edited by Jorge del Real Olvera
p. cm.
Print ISBN 978-953-51-2566-2
Online ISBN 978-953-51-2567-9

INTECH

open science | open minds

Contents

Preface

One important unit operation used in numerous industrial processes is the drying; this is recognized as the leading industrial consumer of energy derived from fossil fuels in many countries. The drying process involves reduction of moisture content contained in a solid material by removing water under controlled conditions. As standard of living rises in the world, energy usage for drying operations will rise along with the demand for energy efficient, environment-friendly and cost-effective drying technologies, which will continue to increase in the world. Sustainability should be a key component of this process of production, safeguarding resources and reserves for future generations. Moreover, because fuel prices are in constant rising, it is important and necessary for developing sustainable drying technologies to take advantage of the fuel efficiently through innovative ideas.

In fact, the fast growth in energy consumption for drying will increase at a velocity more rapid in the world, in particular, in the large economies. The primary goal of this book is to make useful technical literature available for particularly those who have little access to expensive books and journals. Researchers, students and faculty members can use this book for teaching purposes as well as for technical and industrial needs. Contributors of this book have tried to put their ideas in simple terms without sacrificing quality to cover the state-of-the-art and future directions in sustainable drying technologies.

This interdisciplinary and comprehensive volume, consisting of five chapters, covers a survey of trends in sustainable drying technologies for the development of functional foods, spray drying of juice, thermostability of freeze-dried for plants and evaporative drying of low rank coal. After reading this book, I am certain that you will find justified reasons to start your own personal and social awareness campaign in favour of these effective technologies to science, technology and engineering of drying.

To have a broader picture about sustainable processes, in this book, the first chapter provides a simple and convenient introduction to the basic principles and terminology of the drying, in addition to classification, details on commonly used dryers and new developments in drying.

The second chapter contains detailed information about some measurements taken by the food industry to ensure the supply of essential nutrients to as many individuals as possible assuring the global sustainability. More specifically, the contribution of some drying techniques is employed in the development of functional foods to increase the sustainability of the feeding process.

The powders obtained from drying of fruit juices, honey and lactose are examples of sticky products. These products are difficult to dry in a spray dryer under standard conditions due to high sugar content (sucrose, glucose, lactose and fructose) and organic acids (citric, malic and tartaric acid). With this background, the third chapter describes the effect on the use of atomizers (nozzle and rotatory disc) and drying temperatures during spray drying of *Opuntia xoconostle* juice on microstructural, functional and physicochemical properties of the product obtained.

In the fourth chapter, we talk about the most effective parameters on evaporative coal drying process with data in the recent literature open to the authors. Effective parameters are evaluated in three categories, as follows: (1) parameters about drying media, (2) coal parameters and (3) drying method. Effective parameters about drying media are type of media, temperature, pressure, velocity and relative humidity. Different coals in varying sizes are investigated in the section of parameters about coal. Finally, drying methods used in the literature are studied. The main aim of this chapter is to summarize the recent studies on LRC drying and to investigate most effective parameters on drying.

Finally, in the last chapter, we make a description of the analysed and optimized stages of the sation lyophilization parameters and material formulation, based and heavily confronted with current state of knowledge, including so far rare reports of sation lyophilization of plant materials containing other antigens, both at the initial stage and during further research. We present a comprehensive view on the development process and main problems, which we faced during our work regarding lyophilization, associated processes and reproducibility factor. During establishing parameters of drying process, different proteins require unique approaches. This chapter should give a significant insight into practical aspect of designing successful parameters for material that is not widely used but gains attention. Therefore, it should pose a valuable tool and source of knowledge for future freeze-dried bioproducts.

I believe this book is suitable for self-study by engineers and scientists trained in any discipline and so as for the readers who have some technical background. It should also be helpful to industrial users of dryers and dryer manufacturers. The topics chosen are designed to give readers a quick practical overview of the sustainable drying without going into deep mathematical or theoretical considerations. I hope that this book will be useful to researchers working in many drying process.

Dr. Jorge del Real-Olvera
Environmental Technology, Centre of Research and
Assistance in Technology and Design of the State of Jalisco, (CIATEJ),
Guadalajara, Mexico

Chapter 1

Introductory Chapter: Principles of Sustainable Drying

Jorge del Real-Olvera

Additional information is available at the end of the chapter

http://dx.doi.org/10.5772/64614

1. Historical background

Currently in many industrial processes, removal of water or residues of different finished products can pose a serious technical, economic and environmental problem. To solve this, the unit operation of "Drying" is employed. This operation is the world's oldest method for preserving food, elaborating building material, producing pigment and drying animal skins.

In the past centuries, people in the near East kept fruits by wrapping them in dried palm leaves and burying them in hot sand to dry. Moreover, people in the Arctic made deposits with the surplus of walrus meat, "freeze dried" it by piling up stones away from their home to avoid predators from devouring it. Native Americans in the northern parts of the United States had employed the smoke from fire for drying meat, herbs, vegetables or fish through recirculation of hot air.

For many years, the Indians in Peru dried potatoes with a simple but effective process. In the first place, the material was frozen all night outdoors and the next day, the potatoes were fast thawed, with that the moisture content remaining in the material boiled down. Then, they air-dried the potatoes until crisp enough to be stored. Almost everywhere in the world, people have utilised different food-drying methods in order to save food from one season to the next, be it grains, meat, fruits, or herbs among others.

Over time, people in different cultures perfected drying wild and cultivated foods. For example, The Greeks and Romans dried peas and grapes successfully. The Persians learned how to preserve apricots and melons. The Chinese and Japanese were clever at the art of food preservation, cured fish and sea vegetables. Mongolian explorers, en route to Europe, packed bundles of dried milk products to sustain them on their long journey [1].

Across time, the process of drying fish and meat has supplied for many people around the world the proteins necessary to live. In other times, the ancestors dried foods and conserved these for years and years, without the advantage of refrigeration. In fact, the most commonly dried foods were different fish and meat. Ironically the fish is one type of food most difficult to dry, because of its ease for sheltering bacteria in the products without treatment.

Ancestrally, different meats like fishes were treated with curing salt or a brine solution, which helps remove water out of the meat. Finally, fish and meats were treated with smoke for a smoking process [1].

2. Basic theory of drying

The efficient employment of energy in the drying process, the reduction in cost and the minimum impact on the environment are aspects enormously lucrative for the industry. Accordingly, recently a considerable number of reports have been developed around the theme of sustainable drying in both industrial and academic areas. Sustainable drying is intrinsically a multidisciplinary field, which has a very extensive demand in areas such as agriculture, chemical, food, textile, building, tannery and other applications. For understanding this, an optimal knowledge of the three phenomena of transport and suitable knowledge of science of materials are required, because the principal priority is not solely to conserve the energy but, furthermore, to achieve a better product quality through a sustainable drying process [2].

Sometimes it is necessary that products meet some specific requirements after the drying process, like size of the product, free movement or dust free, resistance, particle size distribution, solubility or preservation of active component. As a result, the selection of an adequate process of drying and the design of dryer has a big influence over the quality, shape, size and moisture content of the final product [3].

Because of these, sustainable drying has been a significant part of research since the decade of 1980s. Many studies about sustainable drying generally have two important components: one is the *drying theory*, which treats the analysis of the three phenomenon of transport, and the other one, the *drying equipment*, which has the objectives to design, optimise and fabricate dryers based on the last theory. The drying phenomenon can be defined as a simultaneous process of transfer in momentum, heat and mass, where the end result is the elimination of humidity from the desired products to reduce the water content in them [4].

It is well known that the drying takes place principally through two mechanisms: the movement of humidity from the interior of a material to the outside and the evaporation of humidity from the hot surface of a material to the surrounding. The transport of water are intimately related to some external factors like humidity, pressure, the nature and type of the exposed surface, temperature, and flow velocity [5–7].

Due to its complexity, the investigations on drying are still an area of interest to numerous researchers all over the world. The principal motivation in many of these projects of research in drying is to define the influence of the external factors over the process. The understanding of the drying process with all detail is needed for the design precise of equipment employed scientific principles, maintaining the quality of the product and energy optimization, but overall from a viewpoint sustainable [8–11].

Despite the increase in the number of studies and technological developments, drying process is nowadays an exclusive area to be studied in detail and extensively by technicians and researchers. It is necessary maintain efforts to make the drying process be more clear for all the people; however, this is a complicated physical phenomenon and is not easy to establish the design and control of the equipment and support the product quality. The employ of knowledge in thermodynamics and transport phenomena for the description of equilibria and kinetics drying is between the modern challenges for the research in this field [12].

3. Classification of drying methods and equipment

In general, the drying methods have been arranged and diversified according to the specific necessities of each product. The drying process takes place in different forms and employs diverse types of equipment, depending on the requirements stipulated by the consumer. As seen in **Table 1**, the equipment for drying can be operated fundamentally with two big modes, which are commonly designated as continuous and batch, depending on operation type.

Though the procedure in continuous is most technical and practicable, the dryers in batch are most commonly employed, especially when the control of quality or considerations of health are important factors. This class of process is seen in the pharmaceutical and food industries [12].

The drying in batch is a chemical process and usually preferred when different materials are handled in the same unit of drying, or when the drying is extensive and difficult and in which batch system is placed downstream and upstream, for example, dried wood, brick, among others. The principal difficulty of a system in batch is that the development does not reach a steady state and this causes continuous changes in the operative function [13].

In contrast, a continuous system needs less work, space and energy. This guarantees a supplementary uniformity in the dried product at an inferior cost than a batch system of identical capacity. Sustainable drying has many forms depending on specifications in the products and the use of various types of equipment. Depending on the type of heat applied, drying system involves two basic methods: non-adiabatic and adiabatic. The heat of vaporisation is supplied by the sensible heat in contact with the material to be dried in the adiabatic processes, whereas in non-adiabatic system, evaporation heat is proportionate only by radiant heat or by heat transported through the walls in interaction with the product [12].

The heat applied during the drying can be studied in four different types: convective, radiative, conductive and dielectric. In the process where heat is applied by conductive, heat transfer proceeds if there exists a gradient of temperature in a solid and necessary heat is acquired from a heat source like an electric heater. Usually, direct contact takes place between the heat generator and the target tissues. The heated gas is dispersed into the dryer by means of a fan for help the drying process through evaporation. The word convection is used to describe the transfer of heat energy from some gas or air to liquid, solid or gas. The acquired heat in the gas stream is occupied to evaporate humidity from a surface. The phase used for eliminating the evaporated humidity also makes use of gas stream as a transport.

Judgement	Categorization
Type of operation	Batch Continuous
Type of heat applied	Radiation Convection Conduction Intermittent, continuous dielectric Adiabatic or non-adiabatic
Movement of the material in dryer	Moving Stationary Dispersed Agitated
Pressure of operating	Atmospheric Vacuum
Medium of drying	Air Combustible gases Superheated stream
Temperature of drying	Under boiling temperature Over boiling temperature Lower freezing point
Movement between medium and solids	Mixed flow Counter current Co-flowing
Stages employed	Multiples stages Single stage
Time of residence	Long Medium Short

Table 1. Categorization of drying methods and equipment [3, 12].

Author details

Jorge del Real-Olvera

Address all correspondence to: jdelreal@ciatej.mx

Environmental Technology, Centre of Research and Assistance in Technology and Design of the State of Jalisco (CIATEJ), Guadalajara, Mexico

References

[1] Food Dehydrator Reviews. A Brief History of Food Drying. New York, EU: Fame Themes; 2016. Recuperated in http://dehydratorjudge.com/a-brief-history-of-food-drying/.

[2] Strumillo C. Perspectives on developments in drying. Dry Technol., 2006; 24: 1059–1068. http://dx.doi.org/10.1080/07373930600778056.

[3] Kudra T, Mujumdar AS. Advanced Drying Technologies. New York: Marcel Dekker; 2004. p. 1829–1841.

[4] Toei R. Theoretical fundamentals of drying operation. Dry Technol., 1996; 14(1): 101–182. http://dx.doi.org/10.1080/07373939608917089.

[5] Khraisheh MAM, Cooper TJR, Magee TRA. The transport mechanism of moisture during air-drying processes. Trans IChem E., 1997; 75(C): 34–39. http://dx.doi.org/10.1205/096030897531342.

[6] Telljohann U. Influence of the dryer atmosphere on the danger of drying cracks and dryers scumming. ZI Ziegelind Int/Brick Tile Ind Int., 2003; 3: 12–18

[7] Scherer GW. Theory of drying. J Am Ceram Soc., 1990; 73(1): 3–14. http://dx.doi.org/10.1111/j.1151-2916.1990.tb05082.x.

[8] Land CMV. Drying in the Process Industry. New Jersey: John Wiley & Sons, Inc.; 2012.

[9] Waananen KM, Litchfield JB, Okos MR. Classification of drying models for porous solids. Dry Technol., 2008; 11(1): 1–40. http://dx.doi.org/10.1080/07373939308916801.

[10] Haghi AK, Ghanadzadeh H. A study of thermal drying process. Indian J Chem. Technol. 2005; 12: 654–663.

[11] Kempl C. Progress in dryer selection techniques. Dry Technol., 1999; 17: 1667–1680. http://dx.doi.org/10.1080/07373939908917644.

[12] Yataganbaba, A., Kurtbaş, İ. A scientific approach with bibliometric analysis related to brick and tile drying: A review, Renew Sust Energ Rev., 2016; 59: 206–224. http://dx.doi.org/10.1016/j.rser.2015.12.357.

[13] Parti M, Palancz B. Study of batch drying. Int J Heat Mass Transf., 1973; 17: 669–679. http://dx.doi.org/10.1016/0017-9310(74)90200–2.

Chapter 2

Thermostability of Freeze-Dried Plant-Made VLP-Based Vaccines

Marcin Czyż and Tomasz Pniewski

Additional information is available at the end of the chapter

http://dx.doi.org/10.5772/63503

Abstract

Freeze-drying or lyophilisation is a technique commonly used for pharmaceutical products, in which stability is required to be preserved beyond 4°C temperature. Although lyophilisation is a well-established method, designing a sustainable process for a new product is still to a large extent subjected to empirical practice and often requires trial and error approach. Moreover, even successful lyophilisation of the product may not assure its good long-term storage stability, and progressive decrease in activity may still be encountered. In the past decades, numerous studies have been conducted in the area of protein instability and preservation during lyophilisation and long-term storage. Many critical issues have been identified with regard to physical and chemical instability of proteins in the solid.

Our research effort was focused on designing the prototype oral plant-made vaccine against hepatitis B virus (HBV), which is based on expressing viral surface antigens: small, medium and large (S-, M- and L-HBsAg), assembled into virus-like particles (VLPs) in lettuce plants. We describe in detail the process of developing a freeze-drying protocol facilitating successful processing of plant tissue containing particular HBsAg antigens, while preserving their VLPs structure and immunogenicity. The processing of all HBsAg proteins was investigated, yet the research was focused on S-HBsAg as the basic structural antigen and the main viral immunogen, applied in commonly used preventive anti-HBV vaccines. Each lyophilisation step was consecutively evaluated. Protective excipients, freeze-drying profile, associated processes and long-term storage were investigated to select the most effective conditions. Finally, sucrose demonstrated appropriate properties as a lyoprotectant and the profile of 20°C for 20 h for primary and 22°C for 2 h for secondary drying, as well as nitrogen atmosphere for long-term storage, enabled to maintain S-HBsAg VLP structure and antigenicity.

The preservation of native and immunogenic S-HBsAg in plant-derived preparation was confirmed during mouse immunisation trials when the orally administered preparation was used as boosting and elicited an immune response comparable to routine injection vaccination. Obtained results provided basis for an efficient freeze-drying process that in the future can be utilised for the purpose of a suitable oral plant-derived vaccine against HBV. On the example of anti-HBV oral vaccine, this chapter presents comprehensive view on the development process and main problems, which can be faced during work on freeze-dried products containing bioactive proteins of plant origin. A significant insight into practical aspect of designing successful parameters for material that is not widely used, but gains attention, was given.

Keywords: Freeze-drying, excipient, long-term storage, neutral atmosphere, plant-made oral vaccines, plant tissue, thermostability, HBsAg, VLPs

1. Introduction

The original concept of plant oral vaccines dates back over 20 years. Some of the first and flagship research projects were studies on vaccines against hepatitis B virus (HBV).

To date, the epidemiological situation connected with HBV around the world, particularly in developing countries, has not improved markedly [1, 2]. The first attempts at providing a solution to this problem by oral vaccination were based on the idea of immunisation through consumption of raw tissue of transgenic plants expressing HBV surface antigens (HBsAg). However, despite some encouraging results [3–8], as induction of a specific humoural response [9], it was soon outlined that the future oral vaccine cannot be founded on unprocessed raw plant material [10, 11]. A practical application of fresh plant material in the form of a medication would present a number of problems in both the distribution and the application itself. Limited durability, uneven antigen content throughout the harvested biomass and fostering oral tolerance acquisition were only some of the major forecasted problems. Therefore, the original concept evolved from plant-based 'edible' vaccine to orally delivered lyophilised preparation.

However, freeze-dried formulations facilitate elimination of complex material purification, size reduction and better stability during storage, as well as easy handling and a controlled administration regime, including dosage and short-duration delivery. This was highly attractive with regard to priorities of efficacious, cost-effective, and reliable mass hepatitis B vaccination programmes in developing countries [1, 2, 12, 13]. Preliminary trials confirmed that freeze-dried material containing the small surface antigen of HBV (S-HBsAg) without exogenous adjuvants induced a systemic immune response above the nominal protective titre in mice. Nevertheless, lyophilisation of plant material required further investigation, since 90% degradation of S-HBsAg and other surface antigens—medium and large (M- and L-HBsAg) in the specific immunogenic form assembled into virus-like particles (VLPs)—was observed during that process [10, 11].

2. Basics of protein lyophilisation

Freeze-drying or lyophilisation is widely used to preserve biologically active proteins and polypeptides, including biopharmaceuticals, which are physically and/or chemically unstable in aqueous solutions [14]. While it is a well-established method, commonly used for pharmaceuticals, it continues to be subjected to empirical practice in terms of its optimisation or adaptation to new candidates [15, 16]. This method has great potential for improving the stability of labile substances, especially proteins [17] and liposomes [18, 19], and also generates physicochemical stresses, which can denature proteins to various degrees. In addition, even after successful lyophilisation, the obtained product may still have a limited long-term storage stability. To preserve protein from denaturation caused by freezing (cryoprotection) and/or dehydration (lyoprotection), a stabilising excipient(s) may be used, in parallel to establishing optimal process profile [15, 17, 20].

Freeze-drying is a complex process involving several steps and depending on various factors. Each of them can have a direct or indirect impact on the efficiency of the process, which is defined as preservation of a protein in the biologically active form. In order to achieve high efficacy, both lyophilisation parameters and material formulation must therefore be carefully optimised. In general, a freeze-drying process comprises three stages, namely freezing (material solidification), primary drying (ice sublimation) and secondary drying (moisture desorption). Efficiency of a freeze-drying cycle depends on process variables, such as cooling rate, shelf temperature and duration, combined with the presence of protective additives [16, 17, 20].

A variety of substances are used as effective formulation excipients for their cryoprotective and/or lyoprotective qualities, among which sugars and polyols are the most common [16, 17]. These can serve both as the amorphous phase protecting an active agent and as a bulking component providing desired physical properties of a lyophilised solid, sometimes expressing those abilities in parallel.

Mechanisms of protein preservation in the dry state are still not sufficiently understood to explain and predict the stability after a protein loses its hydration shell [21, 22]. There are two main hypotheses that provide some explanations of protein stabilisation during lyophilisation and subsequent storage in a dehydrated state. The first is the water substitute hypothesis [23–25], and the other is the glass dynamics hypothesis [26]. However, neither sufficiently explains all the observed aspects of freeze-drying.

The water substitute hypothesis assumes that physical denaturation of a protein is inhibited according to the principles of thermodynamics. Preservation of hydrogen bonds maintains a low value of free energy and thus protects the native protein structure. During drying, substances, such as sucrose or polyols, can form hydrogen bonds by their hydroxyl groups at specific locations on the surface of a protein, substituting removed water and stabilising a protein [17, 27]. Thermodynamic stabilisation is ensured by maintaining balance between native conformation and the unfolded structure. A stabiliser increases the free energy of unfolding of a protein and thereby shifts the equilibrium towards its native, more stable state.

The glass dynamics hypothesis assumes that a stabilising agent forms a glassy matrix with very high viscosity, which extremely reduces molecular mobility. This glass is formed by substances, such as sucrose or mannitol, which do not or only partially crystallise during the solution cooling process. The point at which a dissolved substance (liquid state) passes into a glass with a very high viscosity is referred to as glass transition temperature (T_g). However, this conversion is not considered as a physical phase transition [28]. In contrast to crystallisation, the glassy state preserves an interaction between the excipient and the protein, thus allowing limited mobility to be extended to a protein and consequently significantly slowing degradation processes. However, at temperatures above T_g, the system may enter into a more mobile state. If a specimen is maintained in this state for a sufficient time, irreversible degradation may occur especially in the case of complex structures such as proteins. Therefore, glass transition temperature is also referred as collapse temperature (T_g) and during freeze-drying the temperature of a given formulation must be kept below its specific T_g'. Selected excipients may increase collapse transition temperature and thereby enhance stabilisation of a formulation during both freeze-drying and storage [29].

In the first stage of the freeze-drying process, a material must be frozen, which facilitates immobilisation of solution components and prevents foaming during the vacuum stage [30]. It also facilities the formation of a structure of ice crystals in the frozen mass, which will directly determine the rate of subsequent sublimation and the final morphology of a dried product.

Water crystallisation does not occur at the same time throughout its entire volume but progresses from nucleation sites and forms ice crystals. Since water crystallises in the form of pure ice crystals, it causes condensation of the remaining solution. Also, at this stage, an interaction forms between the lyophilised protein and protective excipients. A glassy state is created by the immobilised protein with established direct hydrogen bonds. Also, freezing has a direct impact on drying rate. Creation of large ice crystals, and thus, a highly porous product facilitates easy migration of water vapour from the product and is achieved by relatively slow freezing (0.2–1.0°C/min). Fast freezing produces small ice crystals, hindering the release of vapour from ice and thus limiting heat loss, which leads to overheating and product collapse [21, 31, 32]. However, slow cooling promotes inactivation of dried proteins due to their prolonged exposure to high local concentrations of different chemicals, pH shift, phase separation or cell membrane dehydration as liquid water is transformed into ice crystals [33–35].

With the progress of cooling, the formed material reaches the point, at which its entire volume is solidified. Typical lyophilised formulations are not subject to crystallisation of all components during freezing. Active proteins and added protective substances remain in a non-crystalline, glassy form, which allows them to interact.

Annealing is an optional step often included after freezing and before drying. It is performed by raising the temperature of a formulation after it is completely frozen and maintaining it for a specified period [33]. This step is used for the recrystallisation of all the components that were not completely solidified during freezing. Leaving a pool of ingredients in the glassy state may lead to their spontaneous crystallisation during drying or subsequent storage, which in turn may result in the release of free energy and hence plasticise the glassy phase comprising the excipient and processed protein. However, sometimes the annealing process may also lead to crystallisation of the amorphous protective component, which will result in the elimination of interactions with stabilised protein and its subsequent inactivation [17]. Annealing is also used to convert the ice structure into large crystals, which improves the sublimation rate, resistance to temperature and ensures a more uniform crystalline structure of the entire dried batch.

Drying, the next step of lyophilisation, consists of two stages: primary and secondary drying. The primary stage begins with lowering the pressure in a lyophiliser chamber below the limit of the water triple point, which allows sublimation to start. The optimum pressure range in the chamber is 0.15–0.2 mbar, while fluctuations in this range have little effect on drying efficiency [36, 37]. Lowering pressure below a certain point may even decrease sublimation rate, as high vacuum will isolate the product from the heated shelf and limit heat transfer needed for evaporation [14]. A significant increase in the sublimation rate and shortening of the process duration are achieved almost exclusively by increasing sample temperature. Hence, optimisation of process conditions at this stage focuses mostly on determining a maximally high shelf temperature, but avoiding melting or collapse of a product, leading to the loss of its activity. The end of primary drying is indicated by an asymptotic increase in product temperature up to the temperature of the lyophiliser shelf [38]. At this point, all crystallised ice is removed, and the remaining water content of about 5–30% in the product is in the form of the hydration shelf or water structurally bound to a polypeptide [39].

In contrast to primary drying with a high-intensity vapour flow, the secondary stage is much less effective, removing about 5–10% of total moisture in a sample, but it takes 20–40% time of the whole process. At this step, remaining moisture is removed in response to an elevated shelf temperature and application of intense vacuum [20]. Although the chance of collapse or degradation of a sample is much lower than that for primary drying, it is still possible. Moreover, complex formulations, such as proteins, often require a certain level of residual moisture to ensure good stabilisation of their structure (even up to 5–7%), but in some cases, a very low water content (max. 0.1%) is needed. Therefore, the length of secondary drying is determined by the nature of the product itself [40, 41].

3. Initial lyophilisation trials of plant tissue containing S-, M- and L-HBsAg

The first stage of studies on a plant-derived oral vaccine against HBV was to investigate the stability of viral surface antigens: S-, M- and L-HBsAg contained in lettuce leaves during lyophilisation process. The surface proteins are encoded by a common gene with three autonomous start codons within the same reading frame. Therefore, all proteins contain the common and the largest domain S, which alone consists of S-HBsAg, and carries additional preS domains at its N-terminus: preS2 in the case of M-HBsAg, while L-HBsAg has preS2 and preS1. The S domain/protein once synthesised, dimerise by hydrophobic and disulphide bonds. Dimers are basal structural units of the HBV envelope and they also can self-assembly into VLPs , both naturally and in artificial expression systems. Hydrophilic preS domains are orientated to the outside of the virions or natural VLPs , consisted of S- and other HBs antigens in various proportions. Naturally, M- and L-HBsAg solely do not self-assemble into VLPs , although such recombinant VLPs can be obtained to some extent.

The plant material for lyophilisation was harvested from transgenic plants stably expressing individual HBV antigens and divided into three groups on the basis of their expression level: 'low', 'medium' and 'high'. This categorisation aimed to determine the effect of the initial antigen content, including the behaviour of VLP-formed antigens, on process efficiency. Since the accumulation of particular antigens varied significantly, respective expression groups were different. In the case of S-HBsAg, the 'low' group was <2 μg of the antigen per g fresh weight (FW), 'medium' 2–10 μg/g FW and 'high' >10 μg/g FW, respectively, and for M-HBsAg, the cut-off points were <2, 2–4 and >4 μg/g FW and for L-HBsAg: <1.5, 1.5–3 and >3 μg/g FW, respectively.

Pilot variants comprised untreated material, directly harvested from plants and plant tissue soaked with three excipients, such as sucrose, mannitol and glycerol, at a concentration of 500 mM. Sucrose was selected due to its high efficiency and extensive use in freeze-drying, mannitol as an agent with the potential to form crystals as a rigid structure supporting the glassy phase formed by plant tissue proteins, and glycerol due to its low molar mass and a small molecule size, which could promote its tighter interaction with the tested antigen [13, 16–18, 20, 22, 42, 43]. Freeze-drying was performed at shelf temperatures of 20°C for 20 h for primary drying and 22°C for two hours for secondary drying (profile of 20°C/20h–22°C/2h). Obtained formulations had the desired dry powder form, with the exception when glycerol was used. In that case preparations showed signs of increased viscosity and the tendency to

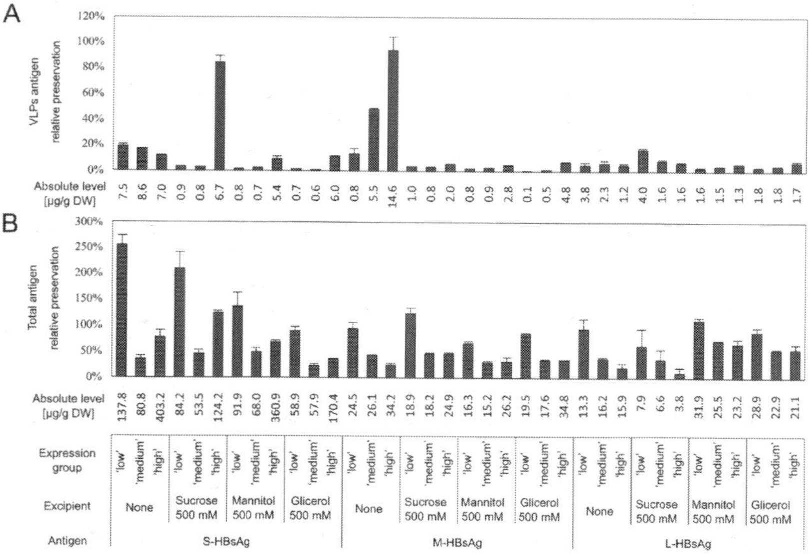

Figure 1. Relative lyophilisation efficiency [%] of lettuce leaf tissue dependent on diverse contents of S-, M- and L-HBsAg in relation to VLPs (A) and total antigen (B). Plant expression groups—'low', 'medium' and 'high' [μg/g FW], respectively, for: S-HBsAg: <2, 2–10, >10; M-HBsAg: <2, 2–4, >4; L-HBsAg: <1.5, 1.5–3 and >3.

http://dx.doi.org/10.5772/63503

clump. The probable reason was the high glycerol content in the tissue after drying, together with its liquid physical state at the tested temperature range of 4–37°C.

Results for individual antigens varied greatly, confirming their different sensitivity to physicochemical factors during freeze-drying. Most likely, despite having a common structural S domain, additional N-terminal preS2 and preS1 domains substantially affect sensitivity of S-, M- and L-HBsAg under stress conditions or their interactions with various excipients. Initial tests have shown low lyophilisation efficiency in most variants—around 10% of preserved VLPs (**Figure 1**), while only exceptionally it was around or more than 20%. However, when the total antigen pool was analysed, a much better efficiency was observed, amounting to 50–100%, which corresponded to several tens to several hundred μg/g dry weight (DW). However, in some samples, a substantial (apparent) increase in the total antigen level was observed. It exceeded 100% when compared to the theoretical full preservation of the antigen after drying.

According to the adopted hypothesis, the increase in the total pool of antigen is associated with the degradation of VLPs formed by HBs antigens. Plant cells express antigens in two pools: one comprises properly assembled VLPs , while the other is composed of antigens forming aggregates and incomplete subparticles and even unstructured 'free' dimers [10]. Effects of various stresses can cause fragmentation of particles, subparticles and aggregates of antigens to dimers, artificially adding to the total amount of antigens. These various forms are in fact indistinguishable in immunoenzyme assays, including commercial tests, utilising monoclonal and polyclonal antibodies specific to the S domain—common for all HBsAg. To some extent, these tests detect 'subparticles', which is reflected as an apparent increase in VLPs contents, also observed by other authors [13, 44]. However, further disintegration of the VLPs is no longer observed as their increase; instead, the total antigen content begins to rise. In addition, only small part of VLPs may be decomposed resulting in a large number of smaller aggregates or dimers, substantially increasing the pool of the 'free' antigen compared to the diminished or even constant quantity of VLPs . At the same time, it must be noted that larger the particle, the higher its susceptibility to decay due to the relatively weaker nature of the constructing bonds. The intensity or nature of a given stress causing VLPs fragmentation may be usually insufficient to denature dimers, which are stabilised by strong hydrophobic interactions and disulphide bonds. However, other stresses can cause degradation of dimers or polypeptide chains themselves, finally reducing the pool of 'free' antigens. Both types of degradation can occur at different times or simultaneously, with the same or different intensity. All these processes are manifested as large fluctuations of VLPs and/or total antigen levels.

Considering that the main functional component of existing and developed vaccines against HBV are VLP-formed by HBs antigens, mostly S-HBsAg, [45], optimisation of the lyophilisation process should focus primarily on the preservation of these structures. However, the level of the 'free' antigen should also show a minimal fluctuation, especially avoiding its growth, due to the possible side effects such as induction of oral tolerance. A small reduction of total antigen content may be acceptable due to the elimination of the unfolded antigen pool while maintaining the VLPs . In the best scenario, after freeze-drying, full preservation of VLPs and an unchanged total antigen level should be observed. However, in view of the higher stability of dimers in comparison to VLPs , a moderate degradation of VLPs with a slight increase in the total antigen level may be reasonably expected [45].

Regarding the potential impact of an initial antigen content on lyophilisation efficiency, HBs antigens showed different effects. Processing of the material with S-HBsAg and soaked with excipients expressed a clear positive trend towards a greater antigen stability together with a higher antigen content. However, differences between the best and the worst results amounted to 81, 8 and 10%, for sucrose, mannitol and glycerol, respectively. An opposite trend was observed for untreated plant tissue, yet dispersion of efficiency was only 6%. The use of sucrose provided good antigen stabilisation in the form of VLPs amounting to 84%, which corresponded to 30.7 μg/g DW. Mannitol and glycerol expressed a significantly lesser protective effect towards VLPs at only 9 and 11%, yet it was still higher in comparison to the variant without any excipients added. Interestingly, soaking of plant material with any excipient from the 'low' and 'medium' groups resulted in a very low freeze-drying efficiency—from 1.1 to – 3.1%, even lower than that of the untreated material, indicating a varied effect of nominally protective substances depending on the type of source material. Preservation of the total antigen pool was much more erratic; however, it showed to be less susceptible to degradation than in the case of VLPs . The use of tissue from the 'low' expression group resulted in the highest total antigen contents for all variants of tissue treatment, whereas for the 'medium' and 'high' groups stabilisation was generally well below 100%. Only when sucrose was used, the total antigen pool increased moderately to 124%, in parallel to the high degree of VLPs conservation.

Freeze-drying of M-HBsAg also demonstrated a positive correlation between VLPs stability and the amount of antigen in the source plant tissue, yet the total antigen and the characteristic preS2 domain were preserved in the reverse pattern [11]. However, unlike S-HBsAg, VLPs preservation in variants with excipients was significantly lower (by up to 7%) than that for untreated tissue (>90% or 14.6 μg/g DW). Yet again, preservation of the total antigen pool showed a reverse trend. An important fact was that freeze-drying of material from the 'low' group resulted in the least altered antigen level.

Preservation of L-HBsAg during freeze-drying showed a distinct pattern in comparison to other HBs antigens. VLPs were very unstable and their preservation in all of the tested variants oscillated around only 10%. The maximum value, 17%—corresponding to 4.0 μg/g DW, was achieved for plant material with a 'low' antigen content, unlike other HBs antigens. The use of sucrose positively affected preservation of this antigen, similarly to S-HBsAg, but distinctly in comparison to M-HBsAg, which is structurally more related. Preservation of the total L-HBsAg and its characteristic preS1 domain was also markedly lower than other HBs antigens, decreasing gradually with an increasing initial antigen content in the plant material [11]. The obtained efficiency ranged from 62–110%, reaching a maximum 16 μg/g DW.

Summarising this stage of research, it was found that some protective substances can increase stability of HBs antigens. Among the tested excipients, a positive effect was observed for sucrose towards S- and L-HBsAg, while mannitol and glycerol showed no significant effect on VLPs stability. The reason for this phenomenon is unknown, but a significant impact of sucrose on freeze-drying efficiency has been reported [46, 47]. Preservation of the total antigen pool using excipients showed significantly larger (30–100%) fluctuations than for VLPs , with no visible advantage of a specific excipient. However, S-HBsAg expressed a similar behaviour of VLPs and total antigen, whereas results for the M and L-HBsAg were much more ambiguous, yet usually a higher VLPs stability was accompanied by a decreasing preservation of total antigen.

Initial tests revealed a distinct stability of VLPs , 'free' dimers and particular antigen domains, depending on the antigen structure and its initial content in plant cells, as well as the physicochemical effect of added protective substances. Observed differences could arise from different requirements for the residual water content after drying or other effects of interactions between molecules of the protective agent and VLPs surface, yet the nature of these processes remains unknown. On the basis of obtained results, it may only be assumed that S-HBsAg retained its VLPs structure due to the interactions with additional substances, most likely by creating hydrogen bonds on the surface of the particle. In turn, the antigen M-HBsAg, due to the presence of the preS2 domain located on the surface of the particles may express a self-stabilising effect and perform the role of a protective excipient in the formation of polar bonds (hydrogen, electrostatic and others) with the surface of the particle. In contrast, L-HBsAg having the largest additional polypeptide chain on the surface of the particle, composed of the preS2 and preS1 domains, may lose the ability to self-stabilise due to the influence of preS1 on the preS2 domain. In addition, a large total size of the two domains may also block access of excipients to the surface of the S domain, which during lyophilisation and water deficit may in effect weaken interactions between the S domains and contribute to the decay of VLPs . A similar effect of blocking the particle by outer domains may be responsible for the low stability of M-HBsAg VLPs in the presence of sucrose. The preS2 domains by binding excipient molecules lose their self-stabilisation ability and additionally isolate the S domain from the excipient. Somehow correspondingly to this hypothesis, in the case of lyophilisation of total antigen and the preS domains, the dominant tendency was connected with their decreasing preservation, reversely correlated with their increasing content in the source material. This phenomenon could be attributed to the limited amount of the 'free' antigen, which can be stabilised by the excipient or residual water, and/or innate properties of a particular HBsAg.

Obtained results implied that further work on the optimisation of the lyophilisation process, as well as efforts to increase stability of HBsAg antigens must focus on preserving VLPs as their more sensitive form, which constitute also the main vaccine immunogen.

4. Preliminary studies on storage stability of freeze-dried S-, M- and L-HBsAg

Preparations of freeze-dried plant tissue-containing S-, M- and L-HBsAg, obtained during the previous step, were subsequently analysed for the stability of the antigens during 3-month storage at temperatures of 4, 22 and 37°C. All variants showed significant fluctuations in terms of VLPs and total antigen stability (**Figures 2–4**).

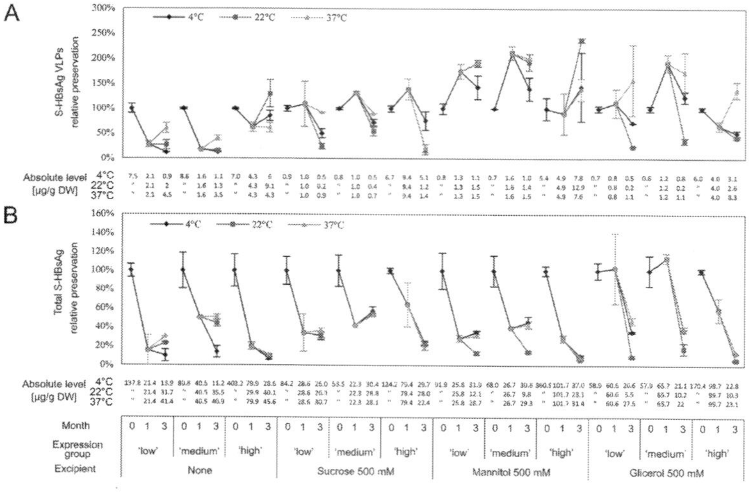

Figure 2. Relative changes [%] of S-HBsAg VLPs (A) and total antigen (B) in lyophilisates derived from tissues of diverse antigen expression levels, then soaked with protective substances and stored at different temperatures. Antigen expression groups, respectively 'low', 'medium' and 'high': <2, 2–10 and >10 µg/g FW.

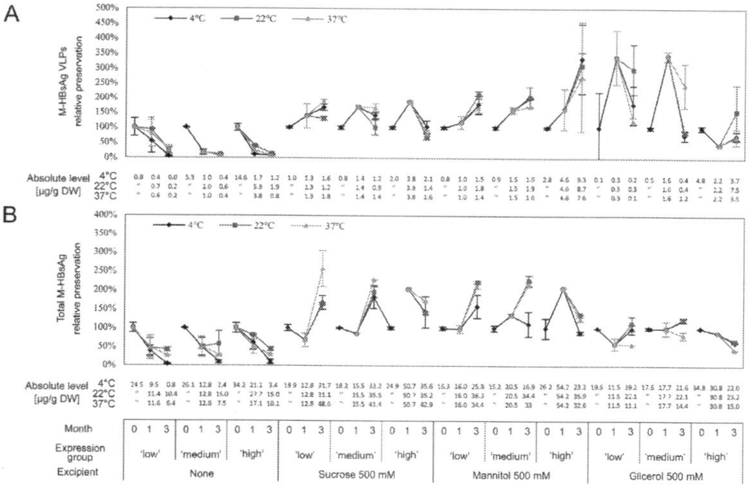

Figure 3. Relative changes [%] of M-HBsAg VLPs (A) and total antigen (B) in lyophilisates derived from tissues of diverse antigen expression levels, then soaked with protective substances and stored at different temperatures. Antigen expression groups, respectively 'low', 'medium' and 'high': <2, 2–4 and >4 µg/g FW. Total M-HBsAg assayed using the ELISA test specific to the preS2 domain.

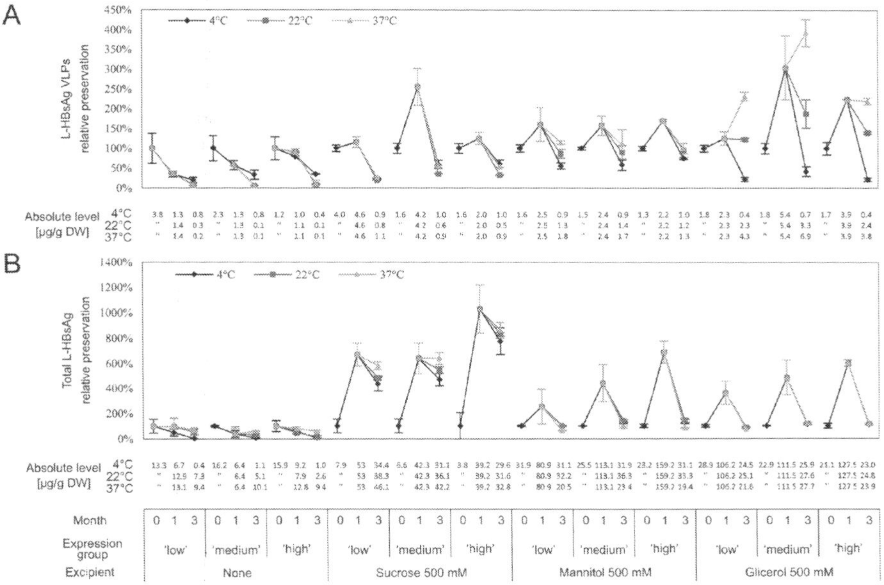

Figure 4. Relative changes [%] of L-HBsAg VLPs (A) and total antigen (B) in lyophilisates derived from tissues of diverse antigen expression levels, then soaked with protective substances and stored at different temperatures. Antigen expression groups, respectively 'low', 'medium' and 'high': <1.5, 1.5–3 and >3 μg/g FW. Total L-HBsAg assayed using the ELISA test specific to the preS1 domain.

During storage of freeze-dried material containing S-HBsAg, in most variants, large fluctuations of VLPs contents were observed, with both a significant decline and an apparent increase (**Figure 2A**), especially at the intermediate point of the period. This should be considered as a negative phenomenon—an indication of VLPs disintegration into smaller 'subparticles' or aggregates. In this context, the smallest fluctuations of the VLPs level were observed in the formulation derived from the material with a 'high' initial S-HBsAg content, soaked with sucrose and stored at 4°C. At higher temperatures, S-HBsAg VLPs were significantly less stable. Total antigen, comprising particles and the 'free' antigen, released during their gradual disintegration, expressed heavy degradation at all temperatures, especially greater than 4°C (**Figure 2B**). However, again, the relatively most stable storage variant of total antigen employed sucrose as the protective agent. In untreated or mannitol-soaked tissue, the total antigen level gradually decreased. Samples treated with glycerol initially exhibited a decreased rate of antigen degradation, but later the antigen level dropped by more than 60%.

In the case of M- and L-HBsAg (**Figures 3** and **4**), degradation processes were more intensive than that for S-HBsAg, especially regarding L-HBsAg. The apparent increase in the particle contents of both antigens after the first month was significantly greater than for S-HBsAg, clearly indicating a strong VLPs decomposition. Material containing M-HBsAg showed large fluctuations in VLPs contents (**Figure 3A**), both at an intermediate storage point and at its end, suggesting an intense VLPs disintegration and following 'free' antigen degradation. The smallest relative changes in the antigen content were observed when sucrose served as the protective excipient, with minor differences between products coming from different plant expression groups. Variants employing mannitol or glycerol underwent a significant apparent increase in M-HBsAg VLPs levels, with the exception of a relatively stable formulation obtained from plant tissue with a 'high' antigen concentration and soaked with glycerol. In addition, avoidance of any excipient for the freeze-drying process, resulted in a gradual, but extensive VLPs degradation during storage. Similar results were observed for the total antigen pool (**Figure 3B**). With the exception of formulations prepared with the use of glycerol, substantial degradation or apparent increase in the antigen level was observed, which indirectly confirmed severe degradation of VLPs and the occurrence of various stresses acting on the unstructured antigen.

Lyophilised plant tissue containing L-HBsAg showed a significant decline of the VLPs level, when no excipient was used (**Figure 4A**). Preparations containing a protective agent showed a significant apparent increase in antigen particle levels after the first month of storage with their subsequent severe reduction in the second storage period. The final VLPs content was comparable to the initial level only in formulations containing mannitol; however, it was after an apparent high increase at the intermediate stage. The level of total antigen (**Figure 4B**) in the formulations with no excipient gradually degraded until its almost complete degradation was reached. In contrast, a significant apparent increase—up to 600% of the initial value, was observed in the formulations containing sucrose. Formulations with mannitol or glycerol showed a significant increase of total antigen after 1-month storage, followed by a decrease to a value close to the initial. This clearly indicated disintegration of VLPs in the initial period, and subsequently, small aggregates and dimers were formed. Variants untreated with protective substances showed only progressive degradation of the 'free' antigen during storage. Nevertheless, at this stage of the study, it could be concluded that excipients had a positive effect on the stability of freeze-dried M- and L-HBsAg, when compared to a significant decrease in their contents in untreated tissue.

Results of preliminary storage experiments indicated a definite necessity to optimise the freeze-drying process as essential for the stability of HBs antigens in derived formulations. Although all antigens, both VLPs and 'free' forms, were gradually degraded, a clear beneficial effect of used protective excipients for their stability was observed (**Figures 2–4**). Additionally, preliminary results showed some important regularities (type of source material, excipients) and phenomena (durability of individual antigens and preS domains, VLPs disintegration), which helped to define directions for the next steps. Among the antigens, S-HBsAg showed the highest stability. This antigen is formed solely by the S domain, which is common, and the main structural element of M- and L-HBsAg antigens, even though the structure of those is heavily influenced by the preS2 and preS1domains. Thus, examination and optimisation of freeze-drying and storage conditions for S-HBsAg could provide a basis for analogous studies on the other antigens. These facts along with the fundamental role of S-HBsAg for current and potential oral anti-HBV vaccines provided a justification for focusing our research mainly on this antigen.

5. Stability of S-HBsAg during plant tissue freezing

An essential step during the lyophilisation process is connected with material solidification by freezing. This often results in a strong degradation of processed proteins [48] or liposomes [49], thus has to be optimised to avoid an unwanted decrease in the contents of active components. The most common approach to alleviate freezing damage is to control the freezing rate of a product. Both fast and slow freezing rates were reported to have the potential of decreasing protein recovery following specific stresses of cryoconcentration, phase separation and cold denaturation characteristic of slow freezing and the formation of a large ice-aqueous interface during rapid cooling [50].

Limited cryoconcentration and phase separation are achieved mainly by optimising product composition and avoiding content reduction for certain substances such as NaCl or PVP. Susceptibility to cold denaturation is an innate trait of a given protein, but certain additives, for example, sucrose, can improve its resistance to damage. Degradation caused by the large ice interface may be prevented mainly by avoiding fast freezing or, if it is unfavourable, by directed ice seeding. This approach, by inducing ice formation, at higher temperatures can effectively slow down the ice growth process and as a result generates a smaller number of larger ice crystals. In the plant tissue, nucleation of ice is observed at temperatures ranging from 0 to -12°C depending on the species, tissue type and its physiological state [51, 52]. However, the ability and effectiveness of modifying this process in plant tissue are very limited and the methods to achieve this are still far from being applied on a regular basis [53]. In addition, inducing ice nucleation outside tissue cannot be effective, due to the membrane and cell wall barrier for ice propagation. Addition of bacteria or chemicals can promote nucleation inside tissue, but this is in contradiction with the principles applied for pharmaceutical formulations and would be difficult to standardise and scale [54].

For these reasons, none of the described methods could be used in the course of the presented work. It was only possible to analyse the effect of the freezing rate on S-HBsAg stability. The research was initiated by examining the effect of plant tissue freezing rate on the integrity of S-HBsAg, both VLPs and the total antigen pool. For this purpose, plant tissue underwent two standard freezing protocols, widely used on both the laboratory and the industrial scales: the 'fast' method utilising liquid nitrogen and 'slow' with a cooling rate of approximately 2°C/min performed in a cooling chamber. Frozen samples were also subjected to 'slow' and 'fast' thawing (water or air bath) to determine its possible impact on the antigen.

In both variants, a significant decrease in S-HBsAg VLPs together with an apparent increase in total antigen contents was observed (**Table 1**). A greater VLPs pool was maintained when the 'slow' freezing protocol was employed: 36–40%, as compared to 16–19% for 'fast' freezing. At the same time, a lower apparent increase in the total antigen was also observed during 'slow' freezing (12 and 21% when compared to approximately 34–36%). This was most probably associated with a lesser extent of VLPs collapse, while the formation of a large area of the ice–aqueous interface during rapid freezing was probably the main cause for VLPs decay. Due to the very high complexity of plant tissue as a processed material, VLPs could also be damaged by ice crystals. This supposition can attest to the fact that the slow cooling protocol, which extended the impact of thickening substances on the antigen, provided a less degrading effect despite the large amounts and variety of compounds in plant tissue. The impact of thawing rate, however, was much smaller—differences between 'fast' and 'slow' thawing amounted to approximately 4% for VLPs and up to 9% for the total antigen pool. Still, for both freezing methods, subjecting the material to 'slow' thawing resulted in a greater stability of VLPs and a less apparent increase of 'free' antigen levels. Since the main functional component of anti-HBV vaccines is the VLP-formed antigen, and a greater elevation of 'total' antigen may cause undesirable side effects such as oral tolerance acquisition, 'slow' freezing and thawing were selected for future work.

Tissue treatment		S-HBsAg			
		VLPs		Total	
Freeze	**Thaw**	**Content [µg/g FW]**	**Change [%]**	**Content [µg/g FW]**	**Change [%]**
'fast'	**'fast'**	3.4 ± 1.1	19.2 ± 6.3	39.5 ± 1.5	134.1 ± 5.2
'fast'	**'slow'**	2.9 ± 1.1	16.5 ± 5.9	40.0 ± 1.2	135.8 ± 4.0
'slow'	**'fast'**	7.0 ± 1.4	39.5 ± 8.0	33.0 ± 2.0	112.0 ± 6.8
'slow'	**'slow'**	6.3 ± 1.7	35.8 ± 9.6	35.6 ± 3.4	120.9 ± 11.7
untreated		17.7 ± 1.6	100.0 ± 8.9	29.5 ± 1.3	100.0 ± 4.5

Material treatment: a/freezing: 'fast'—liquid nitrogen, 'slow'—cooling 2°C/min; b/thawing: 'fast'—immersing in 4°C water, 'slow'—placing in 4°C air. Percentage values represent amounts of S-HBsAg compared to untreated reference sample.

Table 1. S-HBsAg contents in transgenic lettuce leaves after the freeze-thawing cycle [46].

6. Optimisation of the lyophilisation process of plant tissue containing S-HBsAg

Previously obtained results [10, 11] and the present experiments on freeze-drying and storage of plant-associated S-, M- and L-HBsAg demonstrated the need to optimise lyophilisation. Among tested antigens, the native structure of VLPs was most effectively preserved for S-HBsAg, together with a smaller increase in the total antigen pool, which also includes products of VLPs decomposition. Utilisation of protective excipients, especially sucrose, significantly increased stability of the antigen (**Figures 2–4**). Further optimisation was aimed at determining the most effective freeze-drying conditions comprising a protective additive and physical parameters to ensure the highest stability of S-HBsAg VLPs during the process and the subsequent storage.

6.1. Stability of S-HBsAg during lyophilisation in the presence of protective substances

The next stage of research was to study in detail pre-selected substances in terms of stabilising S-HBsAg in order to determine the main/basic excipient—which could be used alone or in

http://dx.doi.org/10.5772/63503

combination with other protective agents. In addition to the previously applied sucrose, mannitol and glycerol, the set of excipients was extended to include glycine and glucose, all used in a wider range of concentrations—100, 250 and 500 mM. This selection was performed on the basis of the reported protection efficacy [13, 16–18, 20, 22, 42, 43] and taking into consideration requirements for the oral route of formulation delivery, that is, harmless ingestion and low cost for product scalability. Glycine was considered as a good crystalline substance with an ability to stabilise the microenvironment of lyophilised proteins [20, 42]. In contrast, glucose, despite the frequently occurring adverse Maillard reaction, was used because of its high ability to penetrate and adhere to protein surface [17]. Initially, two drying profiles were used—20-h primary drying at 5 or 20°C, combined with secondary drying at 22°C for 2 h, identical for the two variants (in short: 5°C/20h–22°C/2h and 20°C/20h–22°C/2h).

The experiments showed that drying at a higher temperature resulted in a much higher degree of VLPs conservation in all excipient variants and for untreated tissue (**Figure 5**). For most of the added substances, a relatively low preservation of VLPs was observed in parallel with large fluctuations in the total antigen pool. However, considerable protective properties were expressed by sucrose and mannitol at concentrations of 250 and 500 mM. Moreover, good stability of VLPs was accompanied by moderate changes in the total antigen level. In particular, the variant with tissue saturated with 500 mM sucrose solution and freeze-dried at 20°C was significantly more efficient than the other options. However, a rise by 24% was observed in the level of VLPs above the baseline level, which most likely resulted from partial dissociation

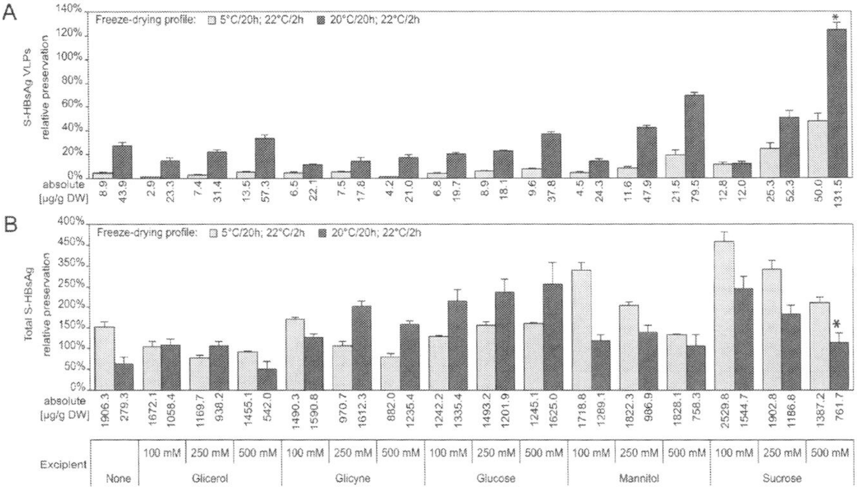

Figure 5. Efficiency of freeze-drying represented as preservation of S-HBsAg VLPs (A) and total antigen pool (B) in lettuce leaves soaked with a series of excipients and processed under two temperature profiles. Preservation effectiveness [%] was calculated as the ratio of VLP-formed and total S-HBsAg in the powdered lyophilised product to the antigen content in fresh tissue multiplied by weight loss degree. The asterisk indicates the variant selected for further experiments [46].

of native particles into smaller 'subparticles' or antigen aggregates. An additional positive effect was the maintenance of the total antigen pool with only small and statistically insignificant changes. This allowed us to draw a conclusion that VLPs disintegration occurred to a relatively small extent and consequently, sucrose was identified as the most promising excipient for further studies. Its use made it possible to achieve significantly higher final absolute antigen contents in the form of VLPs as compared to previous works—approximately 131.5 μg/g DW versus 11–12 μg/g DW [10].

In addition to the determination of S-HBsAg protection effectiveness of individual excipients, a clear trend of an increasing stability of VLPs with increasing concentrations of used substance was observed. However, it was opposed to the behaviour of total antigen, especially in the case of sucrose and mannitol. Because the preferred vaccine formulation would provide a smaller increase in the 'free' antigen pool, the effect was considered favourable.

To summarise this stage of the study, among the analysed protective additives, sucrose proved to be the most effective for the preservation of S-HBsAg VLPs , and at the same time, its use was beneficial by maintaining the total antigen level mostly unchanged. Hence, this compound was selected as the basis for further studies.

6.2. Optimisation of temperature profile

The successive step was focused on determining shelf temperature and duration of freeze-drying, which are of fundamental importance for process economy and scaling up. A higher drying temperature makes it possible to shorten cycle duration and consequently provides a more cost-effective process; however, the material cannot be overheated, as collapse may occur [16, 17]. Also after-process residual moisture is an important factor, having a significant influence on protein recovery.

Results obtained so far indicated that stability of S-HBsAg VLPs was significantly higher when the profile with 20°C for the primary drying temperature was used. Research at this stage focused on attempts to establish a precise drying temperature and time, by the gradual increase in these parameters for primary and secondary drying. Repeated tests of six drying profiles were conducted and they were combined with an analysis of post-process residual moisture (RM) (**Figure 6**).

Freeze-drying with the primary stage conducted at 5°C confirmed a much lower VLPs preservation—approximately 30%, in comparison to processing at 20°C—with an average 97% of retained VLPs in a series of several repetitions. This may have been connected with a high residual moisture in the preparation obtained after drying at a lower temperature—4.1 versus 2.6%. Extension of primary drying to 36 h with an unchanged temperature (5°C) resulted in only slightly decreased RM (3.8%). Although some increase in VLPs stability was visible, it was still insignificant and connected with an enlarged pool of total antigen. In turn, when prolonged primary drying, that is, 36 h at 20°C, was applied and RM was reduced to only 0.6%, it resulted in a significant drop of VLPs and total S-HBsAg. This further confirmed the importance of adequate RM for antigen stability. Next, it was assessed whether shelf temperature beyond 20 and 22°C can be applied. When secondary drying was conducted at 25°C,

http://dx.doi.org/10.5772/63503

2.5% RM was obtained, but VLPs were preserved with a much lower efficiency. In addition, a sharp drop of the total S-HBsAg occurred, probably as a result of the polypeptide chain damage on the VLPs surface. Increasing the temperature of primary drying to 25°C also yielded poor recovery of total and VLP-formed antigen, which could be the result of excessive drying conditions and too low RM. Eventually, the initial profile proved to be the most efficient, indicating that drying temperatures cannot exceed 20 and 22°C, while final RM should be approximately 2.6%. This profile made it possible to save around 100% of VLPs with a relatively moderate increase in the total antigen content.

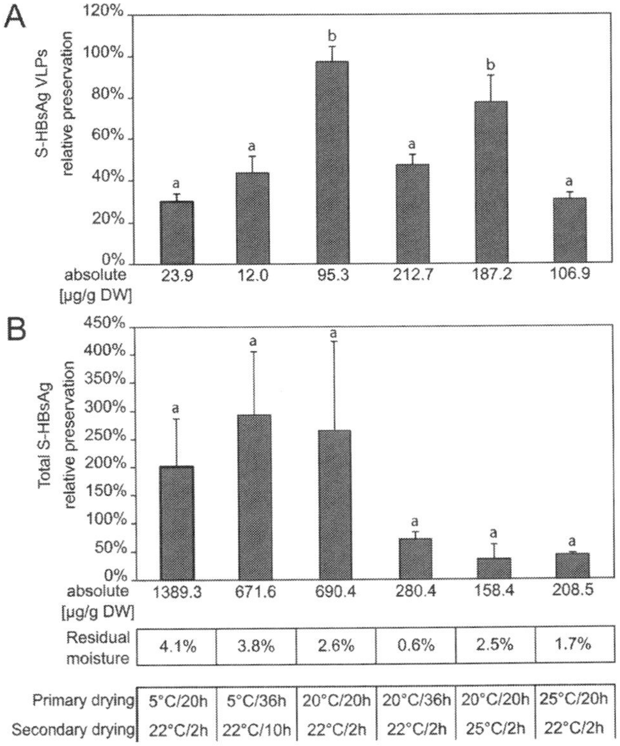

Figure 6. Preservation of VLP-formed (A) and total (B) S-HBsAg in plant tissue lyophilised under different temperature profiles. Sucrose at 500 mM was used as excipient in all variants. Letter indexes mark statistically homogenous groups, separately for the total and VLP-assembled S-HBsAg [46].

In summary, obtained results clearly indicated that for the stabilisation of the appropriate (native) structure of S-HBsAg, both the VLPs and dimers, a certain amount of bound water is required. The water content in the final formulation cannot be too high, since this may favour decomposition processes by enzymatic or physicochemical routes. However, water plays an important structural role. Thus, its content cannot fall below a specific critical value, as this can cause VLPs dissociation and ultimately, degradation of the separate antigen molecules due to the complete loss of the hydration shell. Maintenance of an optimum moisture content at 2.5–2.6% was necessary and possible to establish by maintaining an appropriate temperature and time of lyophilisation.

6.3. Reproducibility of lyophilisation process and batch-to-batch variation

In previous experiments, it was found that the lyophilisation process with the highest efficiency expressed as the highest degree of VLPs conservation, and simultaneously, the smallest shift in the total antigen pool required 500 mM sucrose as the protective excipient and the drying profile of 20°C/22h–22°C/2h. However, the lyophilisation process is a turn-based method; hence, a batch-to-batch variation does occur [16, 43]. Therefore, in the next part of the study, this aspect was investigated in a series of nine repetitions (I-IX), two of which (V and IX) were divided into separate five independently prepared batches but processed during the same single drying cycle (**Figure 7**). The preservation of VLPs in most cases reached approximately 100%, while only in three batches it was less than 50%. It should be stressed that one of those (IX-3) was one out of five homogenous repetitions tested in a single freeze-drying process. The observed fluctuations confirmed a natural process variation. Nevertheless, overall freeze-drying consistency was considered acceptable, as 13 out of 17 batches resulted in an almost complete VLPs preservation. Overall efficiency for all tested batches was 86%, while it was 97% for batches where VLPs preservation reached minimum 50%.

The process reproducibility in relation to the total antigen pool was much more ambiguous. Observed levels of 'free' antigen after lyophilisation ranged from 98% to as much as 556% as related to the theoretical, complete preservation of S-HBsAg and showed no clear connection with VLPs preservation or degradation. The differences in effectiveness were observed between separate processes and during the single drying cycle as in batch V. In some cases, a significant apparent increase in the total amount of antigen was associated with the breakdown of VLPs , as in batches III or VI. In other cases, despite an approximately 100% preservation of VLPs , the level of total antigen increased dramatically, for example in batch V. A possible explanation of this variation during freeze-drying might be that even a partial destruction of VLPs could result in a high increase in the total antigen content, similarly as in freeze-thawing. Even though VLPs are effectively preserved during lyophilisation, some particles may still disintegrate and the released S-HBsAg dimers add to the total antigen pool. However, relatively mild process conditions caused low degradation of S-HBsAg polypeptide chains and resulted in an increased total antigen level. However, in most cases, both VLP-formed and 'free' antigen, remained relatively stable—around 100% of preservation, in particular in repetitions I and VII–IX. It is worth noting here that the total antigen level never fell below 100% in any batch, indicating that the optimised freeze-drying process ensured inhibition of S-HBsAg degradation processes at the molecular level. Despite fluctuations of the total antigen level, it can be said that the key is to achieve high reproducibility of the process in relation to VLPs as the specific immunogen.

http://dx.doi.org/10.5772/63503

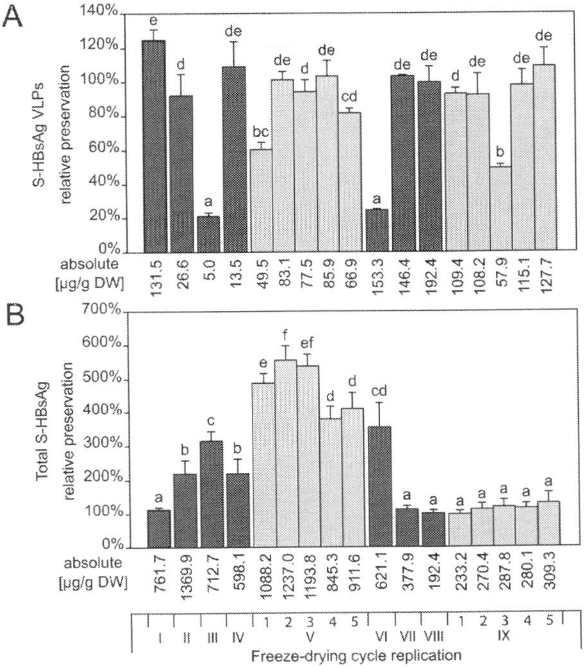

Figure 7. Preservation of S-HBsAg—VLP-formed (A) and total (B)—during freeze-drying replication tests. Each batch was run under the profile of 20°C/22h–22°C/2h and with 500 mM sucrose as the excipient. Batches V and IX were loaded in five separate samples. Letter indexes mark statistically homogenous groups, separately for the total and VLP-assembled S-HBsAg [46].

The absolute content of S-HBsAg in obtained preparations varied significantly, as it directly depended on the antigen expression level in processed plant tissue. VLPs contents ranged from 5.0 to 192.4 µg/g DW, while the content of total antigen in lyophilisate ranged from 196.2 to 1369.9 µg/g DW. However, obtained absolute values of the total S-HBsAg pool were considerably lower than previously reported, amounting to approximately 5000 µg/g DW [10]. Summarising, the elaborated freeze-drying process expressed satisfactory reproducibility regarding S-HBsAg preservation. Furthermore, obtained material was characterised by high VLPs contents along with the total antigen level being not excessive, which is considered beneficial in terms of oral vaccination.

6.4. Freeze-drying-associated processing

Although freeze-drying was the crucial stage in plant tissue processing, the associated handling steps such as tissue infiltration with a lyoprotectant and milling of the product could also affect the final efficiency of the whole process. Hence, both steps had to be closely investigated. Literature data and our results showed that the preservation effect was associated with a high concentration of the protective excipient, here 500 mM sucrose [17, 20] (**Figures 1** and **5**). Therefore, maximal effectiveness of lyoprotectant infiltration had to be ensured. Time (2–15 min) and pressure (50–200 mbar) of soaking were tested (**Figure 8**). The content of S-HBsAg was assayed after soaking and showed no fluctuations in tested parameters. Interestingly, the shortest soaking time proved to be the most effective in all variants, whereas under vacuum of 50 mbar sucrose concentration in tissue was the highest. However, soaking under 100 mbar was only minimally lower and statistically equal. Most likely this lower pressure during soaking created a bigger pressure differential between the chamber and the interior of the leaf tissue, leading to a deeper and more extensive penetration of tissue by the excipient solution. In contrast, prolonged treatment may have resulted in a gradual equalisation of pressure between the chamber and the interior of plant cells, causing greater damage of cells and tissue, and consequently providing a gateway of the excipient. For further work, the variant employing 2-min time and 100 mbar pressure as final parameters was selected due to the milder processing conditions.

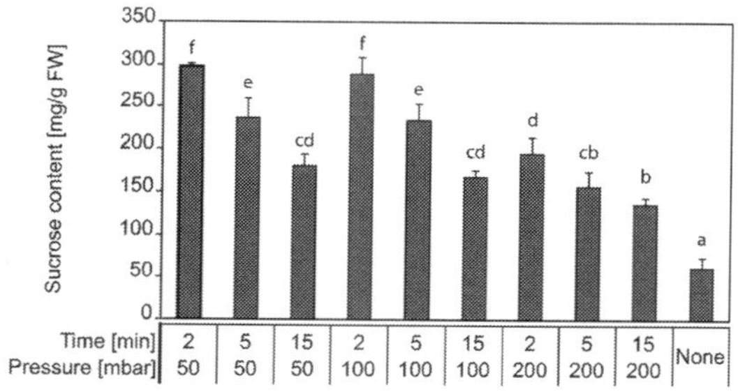

Figure 8. Sucrose concentration after different soaking conditions in lettuce leaf tissue. Letter indexes mark statistically homogenous groups [46].

Lyophilised tissue had to be milled to obtain the final semiproduct. Performed tests showed that within a period of 5–90 seconds, no significant differences occurred regarding both VLPs and total S-HBsAg contents (data not shown). Yet, since approximately 20 seconds were enough to mill the tissue completely, this time was adopted.

7. Storage optimisation of freeze-dried plant material containing S-HBsAg

Initial studies on long-term storage of freeze-dried leaf tissue made it possible to establish that the use of sucrose as an excipient ensures relative stability of VLP-formed S-HBsAg at 4°C, in comparison to other variants [55]. However, VLPs still underwent gradual degradation (approximately 30%). In addition, cold storage is disadvantageous economically and is also in contradiction to the postulated wide availability of a plant-made vaccine. Therefore, research on improving the stability of VLPs in the lyophilised plant material was continued.

A number of physical and chemical pathways of protein degradation may occur during storage in solid states, with almost all being promoted with increasing temperature. The most common and probable pathways comprise a shift of molecular mobility, oxidative and water activity and protein aggregation. Although mobility by principle is limited in the glassy state, on pharmaceutical time scales, it may be sufficient for a considerable molecular mobility of pharmaceutical solids at temperatures up to 50°C below their glass transition temperature [27, 56]. For sucrose, this point is estimated at 60°C in the solid state, and exceeding this threshold during storage at over 10°C could promote collapse of amorphous sucrose [20]. To improve macroscopic mechanical properties of the final product by forming stronger dried cakes and provide support for the amorphous phase, bulking agents, such as mannitol or glycine, may be added. Glycine is widely employed having several advantages, including non-toxicity, high solubility, high eutectic temperature and easy crystallisation [38, 42, 57].

The content of residual water in the product cake throughout its shelf life may be the most important factor responsible for instability [20]. Changes in water reactivity caused by temperature increase, or its content due to transfer from the vial stopper may plasticise the structure of the product cake, thus increasing molecular mobility and promoting various reaction pathways [27, 58]. For this reason, an addition of bulking agents and/or desiccants may have positive effects [44, 59].

Oxidation may be the next major degradation factor for therapeutic proteins and peptides [27, 41, 60]. Despite the complicated and poorly understood theoretical aspects of oxidation processes in dried protein formulations, this degradation can be minimised by sealing vials under neutral atmosphere and/or using chemicals which compete with protein functional groups and bonds for possible oxidisers, for example, reactive oxygen species (ROS). Such antioxidants as sodium sulphite (Na_2SO_3) and ascorbic acid are commonly used as preservatives in the food industry.

In addition to well-known protective agents, several others were also recognised to beneficially affect stability of lyophilised proteins, for instance divalent metal ions, especially zinc [61, 62]. However, mechanisms of their action remain unknown.

Apart from the above-mentioned factors, protein aggregation may have the most essential impact on antigen stability, although reasons of this process are still under investigation [63]. Protein aggregation may take place on physical and/or chemical levels, and it clearly depends on the nature of the protein studied. Currently, the dominant view on aggregation is that the main effort needs to be focused on solutions and rather than trying to prevent protein association, we should avoid or counter conformational alteration in the first place [60]. As aggregation poses a significant threat to freeze-dried formulations, in the case of S-HBsAg, this was not encountered. In fact, disaggregation of VLPs appeared to be the crucial issue in our studies. Therefore, the main effort had to be focused on stabilisation of this antigen form.

Long-term storage tests were carried out on preparations obtained applying the process conditions (soaking, freeze-drying and milling) established previously. Apart from the use of only the basic excipient—500 mM sucrose, it was supplemented with additional substances, which included various stabilisers, for example, divalent metal ions—Zn^{2+}, bulking agents—

glycine, antioxidants—sodium sulphite and/or ascorbic acid, as well as inert atmosphere of nitrogen. All variants tested were used for one year storage at 4, 22 or 37°C in sealed containers, both with and without the presence of silica gel as a desiccant (**Figure 9**).

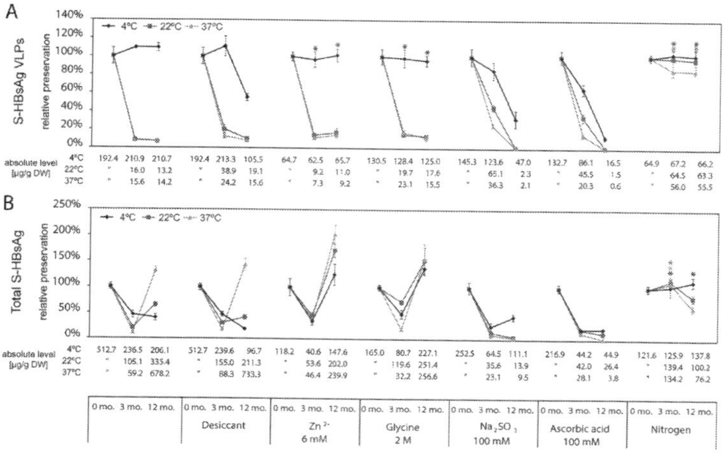

Figure 9. Stability of VLPs (A) and total (B) S-HBsAg in freeze-dried tissue stored for one year at 4, 22 and 37°C. Material was infiltrated with 500 mM sucrose supplemented with different additives and stored under various conditions (with/no desiccant, N_2 atmosphere). Preservation efficiency is represented as a relative change of antigen level to the storage start point. Asterisks indicate insignificant fluctuations in the antigen content with regard to the respective starting point, separately for VLP-assembled and total S-HBsAg.

Most of the tested additional protective substances showed no effect on the efficiency of lyophilisation itself. Preservation of S-HBsAg VLPs and total antigen reached at least 80 and 100%, respectively, which was comparable to the reference material soaked with sucrose alone. Similarly, the moisture content after lyophilisation formulations ranged from 2.6 to 3.0%. Because of the fact that the total amount of the antigen in the lyophilisate directly depended on its content in the initial plant material, storage efficiency was estimated in relation to the starting antigen level immediately after drying, set at 100%. Maintenance of VLPs together with minimal fluctuations of the total antigen level after 3 and 12 months of storage were adopted as determinative in assessing efficacy of the variant.

In almost all tested variants, additional excipients exhibited no enhanced protective effect, since a clear decrease in the VLPs levels was observed, particularly in preparations stored at 22 or 37°C. Total antigen content also fluctuated significantly, indicating that smaller aggregates and dimers released from disassociating VLPs were also subject to severe degradation. In a vast majority of preparations, a degradation scheme was similar to the reference variant, where only sucrose was used.

None of the variants containing zinc ions and/or glycine exhibited a significant improvement in VLPs stabilisation (**Figure 9A**). In turn, the after-process content of the total antigen was higher than in the reference variant, which indicated a reduced degradation (**Figure 9B**). In these variants, total antigen levels initially dropped, yet later dimers and aggregates released from VLPs were degraded to a lesser degree when compared to other variants, hence the final content increased.

Formulations containing antioxidants, that is Na_2SO_3 and/or ascorbic acid, showed initially (for three months) a slower rate of VLPs degradation, both at room temperature and at 37°C. However, at the end of the storage period (12 months), almost the entire pool of VLPs was degraded (**Figure 9A**). 'Free' S-HBsAg also underwent a strong degradation and the final content of total antigen was significantly lower than in the other variants (**Figure 9B**). This effect could result from the pro-oxidant activity of antioxidants, which occurs when these are excessively concentrated. However, the degradation pattern was very similar not only for 100 mM, but also when 10 mM antioxidant solutions were used. This indicated that even a minimal addition of antioxidants may pose an adverse effect on S-HBsAg.

In turn, the use of a desiccant, in parallel with other additives or treatments, resulted in a similar or more intense process of S-HBsAg degradation for both VLPs and the unstructured antigen. A desiccant added to the formulation containing only sucrose as the protective excipient, or additionally kept under nitrogen atmosphere, caused a significant degradation of both VLPs and the 'free' antigen, particularly during the second storage period—between the 3rd and 12th month (**Figure 9**). An adverse effect of the dessicant effect confirmed an irreplaceable role of residual water in the preservation of integrity of both VLPs and the antigen itself.

Only in the variant where the lyophilised tissue was soaked solely with sucrose and stored merely under an inert nitrogen atmosphere, S-HBsAg remained stable during long-term storage. The level of VLPs (**Figure 9A**) was unchanged both at cold and at room temperature. Only at 37°C a small (approximately 15%), statistically significant reduction was observed in the VLPs content. Nonetheless, this occurred only within the first three months of storage, and after that, the level did not change. The content of total antigen (**Figure 9B**) remained completely (statistically) unchanged only when stored in the cold. In contrast, at 22 and 37°C, its initial stable level eventually decreased substantially, to approximately 30 and 50%, respectively. Considering a potential vaccine formulation, this should not pose major problem, as the VLP-assembled S-HBsAg is the immunisation agent. Moreover, when compared to the other variants, and in particular to the reference sample, nitrogen significantly alleviated the degradation process, which suggests a conclusion that oxidative/reductive activities play the most important role in long-term stability of S-HBsAg, as it is for many other freeze-dried proteins [27, 41, 60].

8. Oral immunisation using lyophilised plant tissue-containing S-HBsAg

For the purpose of a prototype plant-derived anti-HBV oral vaccine in the form of freeze-dried plant tissue, apart from maintaining high contents and the native structure of S-HBsAg assembled into VLPs , it was equally important to confirm their immunogenicity. Hence, animal immunisation trials were performed, where mice were vaccinated by intramuscular (*i.m.*) priming with the commercial Engerix®B vaccine and subsequently received an oral booster (*per os, p.o.*) with a suspension of powdered lyophilised tissue based on the previously developed low-dose protocol [10]. The lyophilisate came from batch No. IV with a 109% preservation rate of S-HBsAg VLPs , corresponding to 29 μg/g DW and 199% or 538 μg/g DW of total antigen. Hence, the dose of 50 ng VLPs was delivered in 1.72 mg of freeze-dried tissue per individual subject.

Results of mouse vaccination are summarised in **Figure 10**. The titre of anti-HBs antibodies in mice boosted orally with the lyophilised powder reached a mean value of 293 mIU/ml. Although it was lower, the response was statistically equivalent as in the reference group boosted by an injection with Engerix®B, while the response patterns were also similar. When mice were orally administered the control tissue, no boosting effect was observed. Even though the used freeze-dried preparation exhibited some build-up of total S-HBsAg, apparently it did not hinder immune response development. This result might be obtained due to the considerably lower absolute level of total antigen in the lyophilised tissue than that used previously [10]; however, an exclusively oral immunisation was then performed. Moreover, although the observed response was lower, it was comparable to other reports on injection-oral vaccination trials where plant-associated S-HBsAg was CTB- or LTB-adjuvanted [5, 64]. In conclusion, induction of systemic immune response presented in our study confirmed that the selected parameters of plant material processing ensure successful preservation of S-HBsAg antigenicity and immunogenicity, fundamental to a potential oral vaccine against HBV.

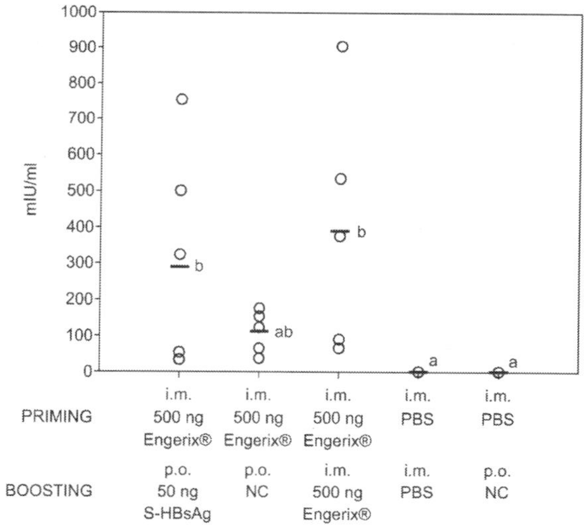

Figure 10. Anti-HBs antibody response in mouse sera after oral boosting with S-HBsAg in powdered lyophilised tissue. Material for immunisation was freeze-dried under the 20°C/22h–22°C/2h profile with 500 mM sucrose as a protective excipient. Symbols: ○ individual mouse response; – group mean value; NC - control lyophilisate (negative control). Letter indexes mark statistically homogenous groups [46].

9. Conclusion

Presented research confirmed that freeze-drying may be exploited for effective processing of such complex preparations as plant tissues containing antigenic proteins assembled into virus-like particles. Here, we showed consecutive steps of our study on HBs antigens to finally establish a technology for S-HBsAg—the basic antigen of hepatitis B virus. The freeze-drying technology, together with storage under inert nitrogen atmosphere, made it possible to obtain a durable intermediate, having a stable content of immunogenic S-HBsAg VLPs , which could be used in the final formulation—a tablet, capsule, etc.—of an oral vaccine against HBV (**Figure 11**). Although adaptation of this method to a specific protein requires numerous steps and repetitions, it appears fundamentally important for the conversion of plant tissue to a form of an oral vaccine against HBV and most probably also other pathogens.

Figure 11. Conversion of lettuce leaf tissue into lyophilised and powdered intermediate with stable S-HBsAg content for preparations of an oral vaccine against HBV.

Author details

Marcin Czyż and Tomasz Pniewski*

*Address all correspondence to: tpni@igr.poznan.pl

Institute of Plant Genetics Polish Academy of Sciences, Poznań, Poland

References

[1] Kew, M.C., Epidemiology of chronic hepatitis B virus infection, hepatocellular carcinoma, and hepatitis B virus-induced hepatocellular carcinoma. Pathol Biol (Paris), 2010. 58(4): p. 273–7.

[2] Romano, L., et al., The worldwide impact of vaccination on the control and protection of viral hepatitis B. Dig Liver Dis, 2011. 43 Suppl 1: p. S2–7.

[3] Thanavala, Y., et al., Immunogenicity of transgenic plant-derived hepatitis B surface antigen. Proc Natl Acad Sci U S A, 1995. 92(8): p. 3358–61.

[4] Richter, L.J., et al., Production of hepatitis B surface antigen in transgenic plants for oral immunization. Nat Biotechnol, 2000. 18(11): p. 1167–71.

[5] Kong, Q., et al., Oral immunization with hepatitis B surface antigen expressed in transgenic plants. Proc Natl Acad Sci U S A, 2001. 98(20): p. 11539–44.

[6] Thanavala, Y., et al., Immunogenicity in humans of an edible vaccine for hepatitis B. Proc Natl Acad Sci U S A, 2005. 102(9): p. 3378–82.

[7] Youm, J.W., et al., Oral immunogenicity of potato-derived HBsAg middle protein in BALB/c mice. Vaccine, 2007. 25(3): p. 577–84.

[8] Neutra, M.R., N.J. Mantis, and J.P. Kraehenbuhl, Collaboration of epithelial cells with organized mucosal lymphoid tissues. Nat Immunol, 2001. 2(11): p. 1004–9.

[9] Kostrzak, A., et al., Oral administration of low doses of plant-based HBsAg induced antigen-specific IgAs and IgGs in mice, without increasing levels of regulatory T cells. Vaccine, 2009. 27(35): p. 4798–807.

[10] Pniewski, T., et al., Low-dose oral immunization with lyophilized tissue of herbicide-resistant lettuce expressing hepatitis B surface antigen for prototype plant-derived vaccine tablet formulation. J Appl Genet, 2011. 52(2): p. 125–36.

[11] Pniewski, T., et al., Plant expression, lyophilisation and storage of HBV medium and large surface antigens for a prototype oral vaccine formulation. Plant Cell Rep, 2012. 31(3): p. 585–95.

[12] Michel, M.L. and P. Tiollais, Hepatitis B vaccines: protective efficacy and therapeutic potential. Pathol Biol (Paris), 2010. 58(4): p. 288–95.

[13] Diminsky, D., et al., Physical, chemical and immunological stability of CHO-derived hepatitis B surface antigen (HBsAg) particles. Vaccine, 1999. 18(1–2): p. 3–17.

[14] Passot, S., et al., Effect of product temperature during primary drying on the long-term stability of lyophilized proteins. Pharm Dev Technol, 2007. 12(6): p. 543–53.

[15] Franks, F., Freeze-drying of bioproducts: putting principles into practice. Eur J Pharm Biopharm, 1998. 45(3): p. 221–9.

[16] Tang, X. and M.J. Pikal, Design of freeze-drying processes for pharmaceuticals: practical advice. Pharm Res, 2004. 21(2): p. 191–200.

[17] Wang, W., Lyophilization and development of solid protein pharmaceuticals. Int J Pharm, 2000. 203(1–2): p. 1–60.

[18] Lang, R., et al., Rational design of a stable, freeze-dried virus-like particle-based vaccine formulation. Drug Dev Ind Pharm, 2009. 35(1): p. 83–97.

[19] Sun, W.Q., et al., Stability of dry liposomes in sugar glasses. Biophys J, 1996. 70(4): p. 1769–76.

[20] Schwegman, J.J., L.M. Hardwick, and M.J. Akers, Practical formulation and process development of freeze-dried products. Pharm Dev Technol, 2005. 10(2): p. 151–73.

[21] Friess, W. and G. Winter, Meeting the challenges in freeze-drying of pharmaceuticals and biologicals. Eur J Pharm Biopharm, 2013. 85(2): p. 161.

[22] Carpenter, J.F., J.H. Crowe, and T. Arakawa, Comparison of solute-induced protein stabilization in aqueous-solution and in the frozen and dried states. J Dairy Sci, 1990. 73(12): p. 3627–36.

[23] Carpenter, J.F., S.J. Prestrelski, and T. Arakawa, Separation of freezing- and drying-induced denaturation of lyophilized proteins using stress-specific stabilization. I. Enzyme activity and calorimetric studies. Arch Biochem Biophys, 1993. 303(2): p. 456–64.

[24] Carpenter, J.F. and J.H. Crowe, An infrared spectroscopic study of the interactions of carbohydrates with dried proteins. Biochemistry, 1989. 28(9): p. 3916–22.

[25] Carpenter, J.F., T. Arakawa, and J.H. Crowe, Interactions of stabilizing additives with proteins during freeze-thawing and freeze-drying. Dev Biol Stand, 1992. 74: p. 225–38; discussion 238–9.

[26] Slade, L. and H. Levine, Beyond water activity: recent advances based on an alternative approach to the assessment of food quality and safety. Crit Rev Food Sci Nutr, 1991. 30(2–3): p. 115–360.

[27] Carpenter, J.F., et al., Rational design of stable lyophilized protein formulations: theory and practice. Pharm Biotechnol, 2002. 13: p. 109–33.

[28] Dobson, C.M., Principles of protein folding, misfolding and aggregation. Semin Cell Dev Biol, 2004. 15(1): p. 3–16.

[29] Chen, M., et al., Immune tolerance split between hepatitis B virus precore and core proteins. J Virol, 2005. 79(5): p. 3016–27.

[30] Willemer, H., Experimental freeze-drying: procedures and equipment. In: Freeze-Drying/Lyophilization of Pharmaceutical and Biological Products, ed. L. Rey, May, J. C. Marcel Dekker, New York, 1999: p. 79–121.

[31] Adams, G.D.J., Technologically challenged –freeze drying damage prevention. Med Lab World, 1996. June: p. 43–44.

[32] Willemer, H., Measurements of temperatures, ice evaporation rates and residual moisture contents in freeze-drying. Dev Biol Stand, 1992. 74: p. 123–34; discussion 135–6.

[33] Adams, G.D.J., Lyophilization of vaccines. In: Vaccine Protocols. ed. A. Robinson, G. H. Farrar, C. N. Wiblin, Humana Press, Totowa, NJ, 1996: p. 167–185.

[34] Grout, B., J. Morris, and M. McLellan, Cryopreservation and the maintenance of cell lines. Trends Biotechnol, 1990. 8(10): p. 293–7.

[35] Pashaiasl, M., et al., Cryopreservation and long-term maintenance of bovine embryo-derived cell lines. Reprod Fertil Dev, 2013. 25(4): p. 707–18.

[36] Pikal, M.J., M.L. Roy, and S. Shah, Mass and heat transfer in vial freeze-drying of pharmaceuticals: role of the vial. J Pharm Sci, 1984. 73(9): p. 1224–37.

[37] Pikal MJ, L.J., Rubber closures as a source of haze in freeze dried parenterals: test methodology for closure evaluation. J Parenter Drug Assoc, 1978. 32(4): p. 162–73.

[38] Akers, M.J., et al., Glycine crystallization during freezing: the effects of salt form, pH, and ionic strength. Pharm Res, 1995. 12(10): p. 1457–61.

[39] Carpenter, J.F., et al., Rational design of stable lyophilized protein formulations: Some practical advice. Pharm Res, 1997. 14(8): p. 969–75.

[40] Kramer, T., et al., A procedure to optimize scale-up for the primary drying phase of lyophilization. J Pharm Sci, 2009. 98(1): p. 307–18.

[41] Chang, L.L. and M.J. Pikal, Mechanisms of protein stabilization in the solid state. J Pharm Sci, 2009. 98(9): p. 2886–908.

[42] Varshney, D.B., et al., Glycine crystallization in frozen and freeze-dried systems: effect of pH and buffer concentration. Pharm Res, 2007. 24(3): p. 593–604.

[43] Liu, J.S., Physical characterization of pharmaceutical formulations in frozen and freeze-dried solid states: techniques and applications in freeze-drying development. Pharm Dev Technol, 2006. 11(1): p. 3–28.

[44] Maa, Y.F., et al., Hepatitis-B surface antigen (HBsAg) powder formulation: process and stability assessment. Curr Drug Deliv, 2007. 4(1): p. 57–67.

[45] Bruss, V., Hepatitis B virus morphogenesis. World J Gastroenterol, 2007. 13(1): p. 65–73.

[46] Czyz, M., et al., Freeze-drying of plant tissue containing HBV surface antigen for the oral vaccine against hepatitis B. Biomed Res Int, 2014. 2014: p. 485689.

[47] Madalinski, K., et al., Antibody responses to preS components after immunization of children with low doses of BioHepB. Vaccine, 2001. 20(1–2): p. 92–7.

[48] Cao, E., et al., Effect of freezing and thawing rates on denaturation of proteins in aqueous solutions. Biotechnol Bioeng, 2003. 82(6): p. 684–90.

[49] van Winden, E.C., W. Zhang, and D.J. Crommelin, Effect of freezing rate on the stability of liposomes during freeze-drying and rehydration. Pharm Res, 1997. 14(9): p. 1151–60.

[50] Rathore, N. and R.S. Rajan, Current perspectives on stability of protein drug products during formulation, fill and finish operations. Biotechnol Prog, 2008. 24(3): p. 504–14.

[51] Gusta, L.V., et al., The effect of water, sugars, and proteins on the pattern of ice nucleation and propagation in acclimated and nonacclimated canola leaves. Plant Physiol, 2004. 135(3): p. 1642–53.

[52] Huang, T., et al., Expression of an insect (Dendroides canadensis) antifreeze protein in Arabidopsis thaliana results in a decrease in plant freezing temperature. Plant Mol Biol, 2002. 50(3): p. 333–44.

[53] Hacker, J. and G. Neuner, Ice propagation in plants visualized at the tissue level by infrared differential thermal analysis (IDTA). Tree Physiol, 2007. 27(12): p. 1661–70.

[54] Morris, G.J. and E. Acton, Controlled ice nucleation in cryopreservation – a review. Cryobiology, 2013. 66(2): p. 85–92.

[55] Pniewski, T., The twenty-year story of a plant-based vaccine against hepatitis B: stagnation or promising prospects? Int J Mol Sci, 2013. 14(1): p. 1978–98.

[56] Hancock, B.C., S.L. Shamblin, and G. Zografi, Molecular mobility of amorphous pharmaceutical solids below their glass transition temperatures. Pharm Res, 1995. 12(6): p. 799–806.

[57] Mattern, M., et al., Formulation of proteins in vacuum-dried glasses. II. Process and storage stability in sugar-free amino acid systems. Pharm Dev Technol, 1999. 4(2): p. 199–208.

[58] Lai, M.C. and E.M. Topp, Solid-state chemical stability of proteins and peptides. J Pharm Sci, 1999. 88(5): p. 489–500.

[59] Maa, Y.F., et al., Influenza vaccine powder formulation development: spray-freeze-drying and stability evaluation. J Pharm Sci, 2004. 93(7): p. 1912–23.

[60] Krishnamurthy, R. and M.C. Manning, The stability factor: importance in formulation development. Curr Pharm Biotechnol, 2002. 3(4): p. 361–71.

[61] Carpenter, J.F., et al., Cryoprotection of phosphofructokinase with organic solutes: characterization of enhanced protection in the presence of divalent cations. Arch Biochem Biophys, 1986. 250(2): p. 505–12.

[62] Carpenter, J.F., L.M. Crowe, and J.H. Crowe, Stabilization of phosphofructokinase with sugars during freeze-drying: characterization of enhanced protection in the presence of divalent cations. Biochim Biophys Acta, 1987. 923(1): p. 109–15.

[63] Costantino, H.R., et al., Deterioration of lyophilized pharmaceutical proteins. Biochemistry (Mosc), 1998. 63(3): p. 357–63.

[64] Hayden, C.A., et al., Bioencapsulation of the hepatitis B surface antigen and its use as an effective oral immunogen. Vaccine, 2012. 30(19): p. 2937–42.

Chapter 3

Sustainable Drying Technologies for the Development of Functional Foods and Preservation of Bioactive Compounds

Ester Betoret, Laura Calabuig-Jiménez,
Cristina Barrera and Marco Dalla Rosa

Additional information is available at the end of the chapter

http://dx.doi.org/10.5772/64191

Abstract

Nowadays, the sustainability of a product, a process or a system is assessed according to three dimensions: environmental, social and economic. Sustainability challenges occur at all stages in the food system from production through processing, distribution and retailing to consumption and waste disposal. The promotion of organic and local food is not the only way to reach the sustainability. There is other possibility that implies to continue the production hegemony. Increasing research is being focused on the development of healthy, quality and safety food products adapted to consumer's needs and more environment-friendly processes, that is, processes consuming energy more efficiently, generating less waste and emitting less greenhouse effect gases. Drying technology is applied in the food industry not only for preservation but also to manufacture foods with certain characteristics. Drying technology operations need to be precisely controlled and optimized in order to produce a good-quality product with the highest level of nutrient retention and flavor together with microbial safety. This chapter contains detailed information about some measurements taken by the food industry to ensure the supply of bioactive nutrients to as many individuals as possible, assuring the global sustainability. More specifically, the contribution of some drying techniques employed in the development of functional foods to increase the sustainability of the feeding process is discussed.

Keywords: sustainability, functional foods, drying, bioactive compounds, structure

1. Introduction

Sustainability means meeting the needs and aspirations of the present without compromising the ability of future generations to meet theirs. As a result of environmental imbalances caused by intensive production and massive use of resources, to achieve food and agricultural sustainability, traditionally, the system has been directed toward promotion of organic and local food, but this is not the only way, as explained in [1]; there is other possibility that implies to continue the production hegemony, emphasizing biotechnology and technological panaceas.

Nowadays, the sustainability of a product, a process or a system is assessed according to three dimensions: environmental, social and economic. Sustainability challenges occur at all stages in the food system from production through processing, distribution and retailing to consumption and waste disposal. The development of a sustainable agri-food system places responsibilities on both the natural and the social sciences [2]. While advances in basic and strategic biological research have greatly expanded, the potential to produce nutritious food in an efficient and environmentally sustainable manner, social and economic factors will determine the uptake and value of this research as well as its future direction [3].

Food processing can be defined as the set of operations which allow manufacturing, preservation and distribution of food products from suitable raw materials. The improvement of the food products is now directed toward ensuring nutritional and specific functional benefits. Regarding the process improvement it is directed to ensure the quality and safety of environment-friendly food products, prepared by optimizing the resources used, minimally affecting or even enhancing their nutritional and beneficial characteristics [4].

Sustainable food production stands at the intersection of several growing needs. First, the needs of consumers for improved food security and safety as well as more sophisticated needs. Second, the quest for economic sustainability of food production based on cost reduction and increased product differentiation. Third, the growing concern for reversing the over-exploitation of natural resources, waste generation and the contribution to climate change [5].

Functional foods are foods that beneficially affect one or more target functions in the body, beyond an adequate nutritional effects, in a way that is relevant to either an improved state of health and wellbeing and/or reduction of risk of disease, and it is consumed as a part of a normal food pattern (not a pill, a capsule or any form of dietary supplement) [6]. Many diseases strictly related with diet and lifestyle concern the society because of their prevalence. Functional foods can help prevent or improve these diseases, thus contributing directly to public health. But the functional effect of a food or food component depends on the active component gaining access to the functional target site. Foods are mostly complex mixtures of macro- and micro-components organized in a structure that can trap active compounds, modulating their release or inhibiting their activity. The selection and development of both appropriate food matrix and technological process, able to maintain the active molecular form until the time of consumption is the key step for the success of a specific functional food [4].

This chapter contains detailed information about some measurements taken by the food industry to ensure the supply of essential nutrients and bioactive compounds to as many individuals as possible assuring the global sustainability. More specifically, the contribution of some drying techniques employed in the development of functional foods to increase the sustainability of the feeding process, is discussed.

2. Drying operation

Drying is an energy-intensive well-studied unit operation in process engineering to reduce moisture content in the food matrix to a level that is safe for storage and transportation, to avoid microbial multiplication, slow down/inactivate microbial activity and the associated product quality deterioration. It involves the removal of water from a wet feedstock by inducing phase changes of water from solid or liquid into a vapor phase via the application of heat (except in the case of osmotic dehydration during which the water is removed without a change in phase by the diffusion of liquid water from solid foods to an osmotic solution through an osmotic pressure difference). The process of drying food materials is extremely complex, involving coupled transient mechanisms of heat, mass and momentum transfer processes accompanied by physical, chemical, structural and phase change transformations [7, 8].

Drying is applied in the food industry not only for preservation but also to manufacture foods with certain characteristics. The nature of the process along with the food structural characteristics results in a very marked effect on the quality characteristics of the final product. There are many different methods of drying food materials, each with their own advantages and disadvantages for particular applications. A vast number of dryer designs reported in the literature are due to the differences in the physical attributes of the product, modes of heat input, operating temperatures and pressures, quality specifications on the dried product and so on [9]. The methods most commonly employed for biotechnological and food products include freeze-drying, spray drying, convective drying, vacuum drying, microwave drying, osmotic drying and combinations thereof (reviewed in [9, 10]). Overall, the quality characteristics of the final product are significantly affected by the process conditions and the way it is conducted. Thus, drying operations need to be precisely controlled and optimized in order to produce a good-quality product with the highest level of nutrient retention. The changes caused to the food properties include discoloring, aroma loss, textural changes, nutritive value and changes in physical appearance and shape [11]. Conditions of drying have a great effect on quality attributes of dried product. For example, higher drying temperature reduces the drying time but may result in poor product quality, heat damage to the surface and higher energy consumption [12]. On the other hand, mild drying conditions with lower temperature may improve the product quality but decrease the drying rate thus drying period is lengthened.

The problems of drying are diverse as various food materials with very diverse physical/chemical properties need to be dried at different scales of production and with very different product quality specifications [13]. The materials preserved by dehydration vary a lot, not only fruits and vegetables to probiotic microorganisms and animal products in the food area, but also other biological materials with important physiological activities, such as human blood cells and insulin.

As described in [4], in most cases, drying involves the application of different temperature conditions (e.g., in the case of freeze-drying, the temperature applied can be −30°C or −80°C, and in the case of other methods such as air drying or spray drying, the temperatures can be 45–80°C or 125–140°C, respectively) that cause irreversible damage due primarily to:

- changes in cellular structures (e.g., cell wall, cell membrane) constituting biological tissues and the induction of changes in key properties responsible for product functionality (e.g., cell membrane permeability, mechanical strength of the wall membrane assembly, etc.).
- changes in the chemical structures responsible for the biological value of nutritious components (e.g., protein, fat). The structural changes also cause changes in the technological functionality that these compounds give to the food to which they belong.
- reactions, mainly oxidation, that decrease the functional value of nutritive compounds (e.g., vitamin, antioxidant).

The major challenge is to remove water from the material in the most efficient way with better product quality, minimal impact on the environment and at the lowest capital and operating costs of the process. Today's increased competition due to globalization, together with the growing consumer demand for better quality products, coupled with the need for eco-friendly and sustainable processes to maintain competitiveness with minimal impact on the environment, will continue to seek innovations in the drying process [9].

3. Strategies to increase the functionality of food products in drying processes

One important part of the sustainability to point out is the minimization of residues on the bioactive compounds recovery from the food waste. During bioactive compounds recovery from food waste, it is common to carry out a drying operation in order to concentrate these ones and use the minimum quantity of solvent. A lot of papers have been written studying the optimal exploitation and revalorization of food waste extracting the maximum quantity of bioactive compounds and minimizing the environmental impact. Some examples of articles/ reviews published are those from [14, 15]. However, in this chapter, we focus on the contribution of functional foods to global sustainability concept. In this way, the principal strategies to increase the functionality of food products during drying as indicated in [9] can be divided into three groups. These strategies can be applied regardless of bioactive compound source either being naturally present in the food matrix or derived from food waste recovery:

1. Addition of ingredients that protect the degradation of bioactive compounds.
2. Creation of structural elements that protect/maintain the functionality of bioactive compounds.

3. Prevention of reactions causing a degradation of bioactive compounds and promotion of those that result in a functional effect.

3.1. Addition of ingredients that protect the degradation of bioactive compounds

As mentioned earlier, drying operation involves removing large amounts of intracellular and extracellular water from food matrices that results in structural and biochemical changes that at the end can affect the functionality of bioactive compounds. The bioactive compounds to be protected vary a lot, from probiotic microorganisms to other important biological compounds such as red blood cells and insulin. As a result, a variety of protectants have been added to the drying media in order to protect the viability of these bioactive compounds. Following this strategy, the researchers aim to not only reduce the degradation of bioactive compounds during drying but also increase their functionality.

Regarding probiotic microorganisms protection during drying, a lot of literature can be found. The probiotic microorganisms are dried in order to extend their viability in dried form or during their incorporation into functional foods. Several works show that properly dried microorganisms remain viable during long-term storage at room temperature [16]. However, the stresses suffered during processing may lead to significant losses in viability and functionality. As explained by Iaconellia et al. [17], the stresses applied on microorganisms by drying processes can be divided into two main categories: the mechanical stresses, mainly localized to the cell membrane, and the intracellular accumulation of reactive oxygen species that causes damage to cell proteins, lipids and nucleic acids. Structural changes can lead to membrane deformation that with fast dehydration-rehydration processes result in membrane permability leading to cell death [18–20]. Moreover, reduced water activity induce phase transitions from crystalline to a gel in cell membrane [21], which may lead to leakage and cell death [22].

A variety of protectants have been added to the drying media before freeze-drying or spray drying to protect the viability of probiotics during dehydration, including skimmed milk powder, whey protein, trehalose, glycerol, betaine, adonitol, sucrose, glucose, lactose and polymers, such as dextran and polyethylene glycol [23]. The beneficial effects of the protectants seem to be related to their protective effect on proteins and cell membranes [24].

As reviewed by Meng et al. [25], drying injuries to the cell depend on probiotic strain, drying method and conditions of processing.

Some examples of new protectants and applications in the area of functional foods developments are described later. In most of the studies, not only the survivability of the probiotic cells is considered but also their functionality is measured in terms of enzyme activity, acid tolerance and hydrophobicity.

The benefit of disaccharide protectants such as cellobiose, lactose and sucrose, for maintaining viability and b-glucosidase activity of *Bifidobacterium infantis* UV16PR during freeze-drying and storage in different food matrices was evaluated [26], concluding that at 10% concentration both trehalose and cellobiose significantly enhanced enzyme activity, viability and acid tolerance.

Resistant starch was found to protect *Lactobacillus plantarum* CIF17AN2 during drying process and could potentially protect it from gastric acid and bile exposures [27]. In the same way, whey protein isolate (WPI) was able to protect *Lactobacillus plantarum* A17 in the encapsulation process. A unique layer-by-layer electrostatic mechanism involved in encapsulation of A17 at pH 7 was found responsible for higher survival of cells [28].

The capability of different fiber preparations to protect the viability and stability of *Lactobacillus rhamnosus* during freeze-drying, storage in freeze-dried form and after formulation into apple juice and chocolate-coated breakfast cereals was studied [29]. The stability of freeze-dried *L. rhamnosus* cells at 20°C was higher in chocolate-coated breakfast cereals in comparison to low-pH apple juice. As in freeze-drying stability, wheat dextrin and polydextrose proved to be better carriers than oat flour in chocolate-coated breakfast cereals. In the development of probiotic chocolate, as reviewed by [30], the lipid fraction of cocoa butter was shown to be protective for bifidobacteria.

Regarding other bioactive compounds, trehalose seems to be the most studied protectant. For example, trehalose has shown to have a protective effect on insulin structure, probably via substitution of hydrogen bonds, while the mild surfactant, sodium deoxycholate, was more protective on the native structure of insulin and, therefore, results in high bioactivity mainly due to resistance to the frozen concentration and interface denaturation in a concentration-dependent manner [31]. Intracellular trehalose has been shown to be necessary for successful stabilization of the membrane during freeze-drying of liposomes and cells [32]. In the same way, trehalose-loaded red blood cells lyophilized in the presence of liposomes demonstrated high survival and low levels of methemoglobin during 10 weeks storage at 4°C in the dry state. A detailed investigation on the liposome size revealed that extruded egg yolk phosphatidyl-choline vesicles with an average diameter of 270 nm are the most effective in inhibiting hemoglobin release. Smaller vesicles could access membrane disruptions and be responsible for membrane repair, which was reflected in reduced hemoglobin leakage [33].

Sometimes, the addition of key ingredients can not only help to reduce the degradation of bioactive compounds but also increase their functionality. It was demonstrated that the addition of a cationic amphiphilically modified dextran could act as excipient in drug delivery nanocarriers of dry power inhalation and significantly increase the drug functionality and its effect [34]. In the same way, a dry powder phage K preparation for oral delivery to control *Staphylococcus aureus* using alginate whey protein microspheres was developed [35]. The results showed that maltose provided the best protection to encapsulated phage K during drying. Both the microsphere size and polymer concentrations in the encapsulation matrix were important factors to determine the degree of protection against stomach acids.

3.2. Creation of structural elements with protective effect

Creation of structural elements with protective effect like encapsulation by using spray drying to create a protective structure, the application of drying operation to form edible films and coatings and the use of vacuum impregnation and its subsequent drying are strategies which can reduce the negative effect of dehydration on biomolecules, protect and even improve the functional value of the food.

3.2.1. *Encapsulation*

Encapsulation is defined as a technology of packaging solids, liquids or gaseous materials in miniature, sealed capsules that can release their contents at controlled rates under specific conditions [36, 37]. The main objective of encapsulation is to protect the core material from adverse environmental conditions such as moisture, heat, oxygen or other extreme conditions. Thereby encapsulation can contribute to increase the shelf life of the product; increase functionality, promoting the controlled liberation of the encapsulated bioactive compound in the target site [38] and keep its properties protecting its bioactive compounds. To extend its shelf life and hence reduce food losses is related with a waste of land, water, energy and several inputs used in production, so any technique effectively reducing these losses will also contribute to the more efficient use of natural resources and therefore sustainability.

Regarding encapsulation technologies, spray drying is an economical, flexible, continuous operation, which produces particles of good quality. For this reason, it is the most widely used microencapsulation technique in the food industry. Encapsulation with spray drying is typically used for the preparation of dry, stable food additives and flavors and to protect functional ingredients such as polyphenols and probiotics [4].

In most of the cases, the capsule is mainly made up of polysaccharides, proteins and their combinations for the microencapsulation of antioxidant components and probiotics. Some polysaccharides such as inulin and polydextrose may act as prebiotic, as they are not hydrolyzed by human digestive enzymes, and have been used to protect probiotic bacteria during spray drying and storage [39].

Recently, food industry by-products have raised considerable interest for their use as encapsulant because of being a sustainable source of material. Chiou and Langrish [40] demonstrated in their study that milled citrus fiber can be used as a replacement for maltodextrin-type carriers to encapsulate hibiscus extract. Also, whey protein is an excellent encapsulating material due to its emulsification, gelation and film-forming properties. Denaturing the whey protein ensures higher tensile property and lower oxygen permeability which protects the probiotic cells from adverse gastrointestinal conditions [41, 42]. Concretely, microencapsulation of *Lactobacillus plantarum* with fructooligosaccharide and denatured whey protein as wall material was found to be most effective in maintaining the viability of bacteria after drying, during storage and in simulated gastric and intestinal conditions [43]. Reconstituted skimmed milk has demonstrated to behave as a protective carrier for improving the survival ratio of lactic acid bacteria (LAB) after spray drying. Such protective effects has been attributed to calcium, which might enhance the heat resistance of LAB cells, and proteins, which lead to a mild temperature variation rate that is beneficial to cell survival [44]. Other example of food-derived protein that is able to protect probiotics from hot temperatures is derived from flaxseed (*Linum usitatissimum L.*) and its mucilages; reference [45] has demonstrated its efficiency as wall materials for microencapsulation by spray drying of *Lactobacillus acidophilus* La-05.

3.2.2. *Edible films and coatings*

Recently, the interest in high-quality food products, increased shelf life and reduced environmental impact has promoted the development of edible and biodegradable polymer films and coatings. Extending shelf life is nowadays one of the main objectives of scientific research and industrial application of edible films and coatings on the surface of several foods.

Use of edible film in multiple food-packaging applications has emerged as an environment-friendly technology with regard to its film-forming properties. An edible film or coating of any material used for enrobing (i.e., coating or wrapping) various foods to extend shelf life of the product that may be eaten together with food or without further removal is also considered [46]. Edible film or coating can control moisture transfer, gases exchange, lipid migration and/or oxidation processes. An edible coating is a thin layer of edible material formed as a coating on a food product, while an edible film is a preformed, thin layer, made of edible material, which once formed can be placed on or between food components [47].

Edible films are obtained from food-grade suspensions and are usually molded as solid sheets onto inert surfaces. They are dried and put into contact with food as wrappings, pouches, capsules, bags or casings through further processing [48]. Biopolymer edible films can be formed via two basic technologies: dry and wet processes. In a dry process, the biopolymer relies on the thermoplastic behavior exhibited by some proteins and polysaccharides at low moisture levels in thermo-compression molding and extrusion. And in wet process, biopolymers are dispersed or solubilized in a film-forming solution (solution casting), and drying steps to make the film matrix [49], solvent removal is required to achieve solid film formation and control its properties [50]. In this case, most of the times, drying operation is applied to form the structure not to obtain dried foods as in aforementioned cases. Those drying operations are generally with air flow at moderate temperatures ranging from 30°C to 60°C, depending on the characteristics of the product. When the edible film is applied in a dehydrate product, drying temperature can be higher; in reference [51], an edible film is applied in an apple snack enriched with fructooligosaccharides and *Lactobacillus plantarum* with methylcellulose, acid citric and sorbitol at different temperatures ranging from 50°C to 140°C during a range of 3–90 minutes.

Edible films and coatings contribute to the revalorization of some industrial by-products which are included in their formulation. This is the case of starch, cellulose and hemicellulose from plant origin, chitosan from crustacean, gums, carrageenan and protein extracted from seaweed, whey protein from the dairy industry, gelatin from slaughterhouses and tanneries, plant-based proteins as soybean and sunflower proteins from oilcakes and keratin from feathers [52–55]. The use of by-products contribute to reduce the waste and hence to increase the sustainability of the process.

In addition, edible films and coatings can act as carriers of functional bioactive compounds as antioxidant and/or with antimicrobial properties, bacteria with probiotics effect or antimicrobial and other components which raise the value of the product by increasing the food's shelf life and protecting its physicochemical properties while maintaining its mechanical integrity and handling characteristics [55–57].

3.2.3. Vacuum impregnation

Vacuum impregnation is a mass transfer operation where a liquid medium is introduced into a solid porous food structure due to pressure gradients created [58, 59]. The liquid amount impregnated into the food matrix depends on the food structure (pore size, distribution, morphology and porosity) and on the vacuum force applied (time and intensity).

Vacuum impregnation can be considered as a useful technology to introduce solutes into the structural food matrix to modify its composition. Generally, it is applied to add bioactive compounds to achieve a technological and/or nutritional functionality [4]. In most cases, this technological operation is used as a pre-treatment for other operations such as frying, drying and freezing due to its effectiveness in reduction of enzymatic and browning reactions, without using antioxidants, by removing oxygen from the food matrix [60]. Vacuum impregnation with a subsequent drying operation is a good combination to obtain stable and enriched functional foods. This operation can also be used to mitigate drying effect by introducing protector compounds as sugars, sugar alcohols and non-reducing sugars in the food matrix. Functional compounds added into the food matrix are more protected from oxygen and other degradation factors than the free functional compound itself; hence functional properties and shelf life are improved, even synergies between some bioactive components can occur and enhance its functionality. It has been demonstrated that bioactive compounds provided by foods can have synergic effect, for example hesperidin is more efficient in combination with ascorbic acid [61]. In reference [62], a probiotic apple snack impregnated with mandarin juice and enriched with *Lactobacillus salivarius* spp. *Salivarius* was developed. The inclusion of the probiotic into a food matrix by vacuum impregnation demonstrated a protection against degradation reactions and at the same time, the new structure could permit the liberation of the bioactive compound in the target site hence improving its functionality.

3.3. Prevention of reactions causing a degradation of bioactive compounds and promotion of those that result in a functional effect

Because of the decrease in the moisture content during drying, most of the nutrients present in the food undergo substantial concentration, thus increasing its nutritional value. However, other more sensitive nutrients are irreversibly transformed and/or destroyed during the dehydration step mainly due to the effect of light, oxygen, heat and the presence of sensitizers. The extent of such changes would depend not only on the processing conditions but also on the sensitivity of each particular compound, their interaction with other food components and the protection conferred by structural matrices, such as cells or microcapsules. From deteriorative reactions occurring during drying of foodstuff, those having a chemical basis are basically oxidation and Maillard reactions. Lipids, vitamins, carotenoids and phenolic compounds are particularly sensitive to oxidation which, in turn, can take place enzymatically or non-enzymatically.

Lipid oxidation leads not only to the development of the typical aroma of many meat products but also to the formation of unpleasant odors and flavors. From a nutritional point of view, oxidation may affect the fatty acid composition and fat quality of meat and fish products. Significant decrease in long-chain polyunsaturated fatty acids (LC-PUFA) was reported during dry-cured ham processing [63]. Also the exposure to light and oxygen during sun drying and controlled oven drying induced a noticeable reduction of the most important ω-3 fatty acids, eicosapentaenoic acid (EPA) and docosahexaenoic acid (DHA), in both lean and fat fishes [64]. In addition, free radicals and peroxides originated during lipid oxidation are closely related to the pathology of some cancers, arteriosclerosis, arthritis, neurodegenerative diseases and the aging process [65]. Moreover, oxidized lipids can react with proteins and other food components and reduce their nutritional quality and safety [66]. Regarding the application of a salting process, brine contact with fish has also been reported to enhance lipid oxidation of the highly unsaturated lipids, which is directly related to the production of off flavor, protein denaturation and texture changes [67]. Specific techniques reported to prevent lipid oxidation in fish oil processing include microencapsulation and the application of natural food additives with antioxidant capacity like rosemary extracts, α-tocopherol or polyphenols from grape pomace [68]. Among simpler technical proposals for reducing lipid oxidation during fish and meat drying, those focused on reducing the exposure to oxygen in the drying chamber and, to a lesser extent, the drying temperature are particularly of interest. To this end, satisfactory results have been reported from vacuum drying and ultrasonic vacuum drying [69], microwave drying [70], ultrasound assisted drying [71], freeze-drying [72] and low-pressure superheated steam drying. Although there is little evidence about the impact of such techniques on the lipid profile of treated products, one intuits that these treatments result in more porous structure entailing greater risk of damage by oxidation during further storage. Also in fruits and vegetables, such techniques have resulted in reduction of pigments, vitamin C, phenolic compounds and other minor ingredients losses due to oxidation.

Carotenoids are natural pigments synthesized by plants and microorganisms. Their importance in human nutrition and health is mainly due to their capability to inhibit oxidative reactions. This property is particularly high in the case of lycopene, closely followed by α-carotene and β-carotene and, to a lesser extent, zeaxanthin [73]. Carotenoids may be free in the lipid phase of the food, forming complexes with proteins, bound to carbohydrates or as fatty acid esters. Carotenoids oxidation can be indirectly catalyzed by lipoxygenase, the enzyme responsible for the peroxides formation from lipid oxidation of unsaturated fatty acids, and results in important color changes and losses in antioxidant activity. Isomerization is also involved in carotenoids loss during food dehydration. Indeed, naturally occurring carotenoids are in all-trans form, which is the most stable chemical form to heat treatments. Thermal treatments applied during food processing promote isomerization of trans-carotenoids to their cis-form, mainly on the 9-cis and 13-cis types; it is not entirely clear whether it adversely affects their ability to scavenge free radicals [73]. As reported in reference [74], 13-cis-β-carotene is formed in carrots as the temperature of the product reaches 60°C, when submitted to hot air drying, or even lower temperature, when applying vacuum drying and low-pressure superheated steam drying. Although the antioxidant activity is unaffected in this case, the conversion of trans-β-carotene in any of its cis-isomers might imply a notable decrease in its activity as vitamin A precursor [75]. Negative effects of isomerization are usually offset by an increase in bioavailability. All-trans forms naturally existing in foods are linear, long and rigid molecules, whereas their cis isomers are shorter molecules that can be more easily solubilized, absorbed and transported at a cellular level [76]. Even the irreversible degradation of carotenes by oxidation could be compensated by this increase in bioavailability [77]. For this purpose, losses during processing should be minimized by using one of the alternatives to aforementioned conventional drying techniques. Preventing the loss of cellular integrity also contributes to diminish the incidence of oxidation, as well as some pretreatments, such as blanching and osmotic dehydration. Blanching benefits are attributed to enzyme inactivation, while osmotic treatments for a short period in the presence of sucrose at 30–40°C have been reported to encourage these phytochemicals generation [78].

Other food components having beneficial health effects due to their high antioxidant and antimicrobial activity, and therefore being susceptible to oxidation, include polyphenols and ascorbic acid. Phenolic compounds and vitamin C are known to prevent free radicals formation and reduce molecular damage on DNA, lipids and proteins, which is directly related to a decrease in the incidence of cancer and coronary diseases. They also play a decisive role in the color and flavor of certain fruits and vegetables. Most of the polyphenols are present in foods as esters, glycosides or polymers, that is, as forms that cannot be absorbed [79]. However, as previously mentioned for other functional compounds, structural and chemical changes taking place during fruits and vegetables drying can contribute in increasing their bioavailability during further consumption. In general, reducing the contact with oxygen in the drying chamber by reducing the drying time reduces losses in phenolic compounds, but due to its greater sensitivity to high temperatures, reducing the vitamin C losses might imply a noticeable decrease in the drying temperature [80]. In spite of these considerations, certain fruits and vegetables show an increase in their ability to scavenge free radicals after drying in adverse conditions [81], which has been explained in terms of the generation of new compounds with higher antioxidant activity as the ones resulting from the Maillard reaction.

The Maillard reaction or non-enzymatic browning reaction is the chemical reaction that occurs between compounds with a primary amine function and compounds with carbonyl groups, which generate different flavors and brown color [82]. This reaction is accelerated under alkaline conditions, intermediate moisture content ($0.55 < aw < 0.75$) and high temperatures, but it is also observed under refrigeration [83]. The type of compounds involved also influences the reaction rate, as well as the presence of certain metals. Logically, meat and fish products, with a particularly high protein content, are most susceptible to experience such reaction. However, by-products of the Maillard reaction have also been found to be less in lysine products, such as fruits and vegetables. Adverse effects associated with this reaction include alteration of the organoleptic properties and decrease of the nutritional value since essential amino acids, mainly lysine, and certain vitamins, such as vitamin K and C, are generally involved. In addition, some of the compounds formed in the Maillard reaction are toxic or mutagenic. This is the case of high carboxymethyl lysine that promotes diabetes and cardiovascular diseases, and some recognized it as a probable human carcinogen compounds, such as acrylamide and hydroxymethylfurfural [82]. On the contrary, melanoidins formed at the last stage of the Maillard reaction are non-digestible compounds having antioxidant and antimicrobial activity against pathogenic microorganisms of the colon [83]. Non-enzymatic browning in foods also includes caramelization reaction, but it involves only sugars or polyhydroxycarboxylic acids and usually requires more drastic conditions. Since the pyrolysis of sugars starts at temperature above 110°C, caramelization reaction in foodstuff drying is not as worrisome as compared to other chemical reactions.

4. Energetic considerations

Drying is probably the most energy-intensive process of the major industrial processes because it consumes large amounts of energy and releases significant amount of carbon oxides to the environment [84]. In an energy-intensive industry like heating or drying, improving energy efficiency by 1% could result in as much as 10% increase in profit [85]. Any small improvement in energy efficiency in food drying process will lead to a sustainable development to global energy perspective.

Condition of drying air has a great effect on the quality attributes of dried product. Thus, one of the key issues of drying technology is to reduce the cost of energy sources to increase the efficiency of drying facilities for good quality of dried products. On the other hand, the design of an energy-intensive system for lower cost and higher efficiency is one of the essential approaches for sustainable development [86].

There are a lot of studies modeling drying operation. Most of the times, the models are directed to analyze heat and mass transfer in order to improve the quality of the final products obtained. With the aim to evaluate the drying operation, there are a lot of studies directed to analyze the energy used during process in order to optimize the drying method and contribute to the sustainability of the process. It is necessary to combine all process variables (drying process, installation design, time, temperature and product characteristics) to minimize energetic and product losses.

Usually, an energy analysis is carried out in most of the studies. The energy analysis is a basic and traditional approach to estimate various energy conversion processes [87]. The energy analysis is based on the first law of thermodynamics, which is expressed as the principle of the conservation of energy. According to Singh [88], energy analysis is useful in quantitative evaluation of energy requirements of energy generating and delivery systems and in the detection of mode and evaluation of energy loss. However, it provides no information about the irreversibility aspects of thermodynamic processes. The energy analysis is unable to distinguish the different qualities of energy such as heat quality which is dependent on the heat source temperature.

The exergy-based analysis and subsequent optimization of drying processes is having a growing interest among the researchers. Exergy is the maximum amount of work obtained from a stream of matter, heat or work when some matter is brought to a state of thermodynamic equilibrium with the common components of natural surroundings by means of reversible processes, and is a measure of the potential of a stream to cause change, as a consequence of not being completely stable relative to the reference environment [89, 90]. The exergetic performance assessments not only distinguish the magnitudes, location and causes of irreversibilities in the plants, but also enables the locations, types and magnitudes of waste emissions and internal losses to be determined [91, 92]. The main objective of exergy analysis of drying systems is to provide a clear picture of the process, to quantify the sources of inefficiency, to distinguish the quality of energy consumption, to select optimal drying conditions and to reduce the environmental impact of drying systems. The exergy analysis is being applied to more and more products. In recent years, some articles have been published combining both energy and exergy calculations in order to have a more completed analysis and sustainability evaluation of the process [93].

5. Conclusions

The development of functional foods can clearly contribute to the global concept of sustainability. The negative effects related to the application of extreme temperatures in drying operations can be minimized by incorporating ingredients that protect structural elements, creating protective structures and avoiding degradation reactions. Management of drying processes in an adequate way can contribute to prevent bioactive compounds losses, maintain and even increase the functionality of dried products.

Acknowledgements

This research was supported by a Marie Curie Intra European Fellowship within the 7th European Community Framework Programme. Authors acknowledge the FPI 2014 programme of the Universitat Politècnica de València.

Author details

Ester Betoret[1*], Laura Calabuig-Jiménez[2], Cristina Barrera[2] and Marco Dalla Rosa[1]

*Address all correspondence to: maria.betoretvalls@unibo.it

1 Department of Agricultural and Food Science and Technology, University of Bologna, Cesena, Italy

2 Institute of Food Engineering for Development, Department of Food Science and Technology, Universitat Politècnica de Valencia, Valencia, Spain

References

[1] Spiertz, H. Food production, crops and sustainability: Restoring confidence in science and technology. Current Opinion in Environmental Sustainability. 2010, 2(5–6), 439–443.

[2] OECD (Organisation for Economic Co-operation and Development). Agriculture and the environment: Lessons learned from a decade of OECD work. 2004. Paris: OECD.

[3] Lowe, P., Phillipson, J. & Lee, R.P. Socio-technical innovation for sustainable food chains: Roles for social science. Trends in Food Science & Technology. 2008, 19, 226–233.

[4] Betoret, E., Betoret, N., Rocculi, P. & Dalla, M. Strategies to improve food functionality : Structure-property relationships on high pressures homogenization, vacuum impregnation and drying technologies. Trends in Food Science & Technology. 2015, 46(1), 1–12.

[5] Fava, F., Zanaroli, G., Vannini, L., Guerzoni, E., Bordoni, A., Viaggi, D., Robertson, J., Waldron, K., Bald, C., Esturo, A., Talens, C., Tueros, I., Cebriàn, M., Sebòk, A., Kuti, T., Broeze, J., Macias, M. & Brendle, H.G. New advances in the integrated management of food processing by-products in Europe: Sustainable explotation of fruit and cereal processing by-products with the production of new food products (NAMASTE EU). New Biotechnology. 2013, 30(6), 647–655.

[6] European Commission. Directorate-General for Research, FP7 cooperation-Food. Functional Foods. 2010, 1–28, Brussels, Belgium.

[7] Sabarez, H.T. Computational modeling of the transport phenomena occurring during convective drying of prunes. Journal of Food Engineering. 2012, 111(2), 279–288.

[8] Sabarez, H.T. Mathematical modeling of the coupled transport phenomena and color development: Finish drying of trellis-dried sultanas. Drying Technologies. 2014, 32, 578–589.

[9] Sabarez, H. Drying of Food Materials. Reference Module in Food Sciences. 2016, 1–10.

[10] Walters, R.H., Bhatnagar, B., Tchessalov, S., Izutsu, K.I., Tsumoto, K. & Ohtake, S. Next generation drying technologies for pharmaceutical applications. Journal of Pharmaceutical Sciences. 2014, 103, 2673–2695.

[11] Quirijns, E.J. Modelling and dynamic optimisation of quality indicator profiles of freeze-dried liposomes by trehalose. Archives of Biochemistry and Biophysics. 2006, 242(1), 240–247.

[12] Ho, J.C., Chou, S.K., Chua, K.J., Mujumdar, A.S. & Hawlader, M.N.A. Analytical study of cyclic temperature drying: Effect on drying kinetics and product quality. Journal of Food Engineering. 2002, 51(1), 65–75.

[13] Mujumdar, A.S. & Wu, Z.H. Thermal drying technologies: New developments and future R&D potential. In: Jangam, S.V., Thorat, B.N. (Eds.), R&D Needs, Challenges and Opportunities for Innovation in Drying Technology. 2010. e-Book.

[14] Galanakis, C. Recovery of high added-value components from food wastes: Conventional, emerging technologies and commercialized applications. Trends in Food Science & Technology. 2012, 26, 68–87.

[15] San Martin, D., Ramos, S. & Zufia, J. Valorisation of food waste to produce new raw materials for animal feed. Food Chemistry. 2016, 198, 68–74.

[16] Zayed, G. & Roos, Y.H. Influence of trehalose and moisture content on survival of *Lactobacillus salivarius* subjected to freezedrying and storage. Process Biochemistry. 2004, 39, 1081–1086.

[17] Iaconellia, C., Lemetaisb, G., Kechaouc, N., Chainc, F., Bermúdez-Humaránc, L.G., Langellac, P., Gervaisa, P. & Beneya, L. Drying process strongly affects probiotic viabilities and functionalities. Journal of Biotechnology. 2015, 214, 17–26.

[18] Schwab, C., Vogel, R. & Gänzle, M.G. Influence of oligosaccharides on the viability and membrane properties of *Lactobacillus reuteri* TMW1.106 during freeze-drying. Cryobiology. 2007, 55(2), 108–114.

[19] Dupont, S., Beney, L., Ritt, J.-F., Lherminier, J., & Gervais, P. Lateral reorganization of plasma membrane is involved in the yeast resistance to severe dehydration. Biochimics Biophysics Acta (BBA) – Biomembranes. 2010, 1798(5), 975–985.

[20] Lemetais, G., Dupont, S., Beney, L. & Gervais, P. Air-drying kinetics affect yeast membrane organization and survival. Applied Microbiology and Biotechnology. 2012, 96(2), 471–480.

[21] Milhaud, J. New insights into water–phospholipid model membrane interactions. Biochimics Biophysics Acta (BBA) – Biomembranes. 2004, 1663(1), 19–51.

[22] Potts, M. Desiccation tolerance: A simple process? Trends in Microbiology. 2001, 9(11), 553–559.

[23] Morgan, C.A., Herman, N., White, P.A. & Vesey, G. Preservation of micro-organisms by drying: A review. Journal of Microbiological Methods. 2006, 66, 183–193.

[24] Leslie, S.B., Israeli, E., Lighthart, B., Crowe, J.H. & Crowe, L.M. Trehalose and sucrose protect both membranes and proteins in intact bacteria during drying. Applied and Environmental Microbiology. 1995, 61, 3592–3597.

[25] Meng, X.C., Stanton, C., Fitzgerald, G.F., Daly, C. & Ross, R.P. Anhydrobiotics: The challenges of drying probiotic cultures. Food Chemistry. 2008, 106, 1406–1416.

[26] Basholli-Salihua, M., Mueller, M., Salar-Behzadi, S., Ungera, F.M. & Vernstein H. Effect of lyoprotectants on β-glucosidase activity and viability of Bifidobacterium infantis

after freeze-drying and storage in milk and low pH juices. LWT-Food Science and Technology. 2014, 57(1), 276–282.

[27] Hongpattarakere, T. & Uraipan, S. Bifidogenic characteristic and protective effect of saba starch on survival of *Lactobacillus plantarum* CIF17AN2 during vacuum-drying and storage. Carbohydrate Polymers. 2015, 117(6), 255–261.

[28] Khem, S., Bansal, V., Small, D.M. & May, B.K. Comparative influence of pH and heat on whey protein isolate in protecting *Lactobacillus plantarum* A17 during spray drying. Food Hydrocolloids. 2016, 54, 162–169.

[29] Saarela, M., Virkajärvi, I., Nohynek, L., Vaari, A. & Mättö, J. Fibres as carriers for *Lactobacillus rhamnosus* during freeze-drying and storage in apple juice and chocolate-coated breakfast cereals. International Journal of Food Microbiology. 2006, 112(2), 171–178.

[30] Konar, N., Toker, O.S., Oba, S. & Sagdic, O. Improving functionality of chocolate: A review on probiotic, prebiotic and/or symbiotic characteristics. Trends in Food Science & Technology. 2016, 49, 35–44.

[31] Zhang, Y., Deng, Y., Wang, X., Xu, J. & Li, Z. Conformational and bioactivity analysis on insulin: Freeze drying TBA/water co-solvent system in the presence of surfactant and sugar. International Journal of Pharmaceutics. 2009, 371(1-2), 71–81.

[32] Crowe, L.M., Crowe, J.H., Rudolph, A., Womersley, C. & Appel, L. Preservation of freeze-dried liposomes by trehalose. Archives of Biochemistry and Biophysics. 1985, 242(1), 240–247.

[33] Kheirolomoom, A., Satpathy, G.R., Török, Z., Banerjee, M., Bali, R., Novaes, R.C., Little, E., Manning, D.M., Dwyre, D.M., Tablin, F., Crowe, J.H. & Tsvetkova, N.M. Phospholipid vesicles increase the survival of freeze-dried human red blood cell. Cryobiology. 2005, 51, 290–305.

[34] Varghese Vadakkan, M., Binil Rai, S.S., Kartha, C.C. & Vinod Kumar, G.S. Cationic, amphiphilic dextran nanomicellar clusters as an excipient for dry powder inhaler formulation. Acta Biomaterialia. 2015, 23, 172–188.

[35] Tang, Z., Huang, X., Sabour, P.M., Chambers, J.R & Wang, Q. Preparation and characterization of dry powder bacteriophage K for intestinal delivery through oral administration. LWT-Food Science and Technology. 2015, 60(1), 263–270.

[36] Vilstrup, P. (Ed.). Microencapsulation of Food Ingredients. 2004. Leatherhead Food International: UK.

[37] Desai, K.G.H. & Jin Park, H. Recent developments in microencapsulation of food ingredients. Drying Technology. 2005, 23(7), 1361–1394.

[38] Shahidi, F. & Han, X.Q. Encapsulation of food ingredients. Critical Reviews in Food Science & Nutrition. 1993, 33(6), 501–547.

[39] Avila-reyes, S.V., Garcia-suarez, F.J., Teresa, M., Martín-González, M.F.S. & Bello-perez, L.A. Protection of *L. rhamnosus* by spray-drying using two prebiotics colloids to enhance the viability. Carbohydrate Polymers. 2014, 102, 423–430.

[40] Chiou, D. & Langrish, T.A.G. Development and characterisation of novel nutraceuticals with spray drying technology. Journal of Food Engineering. 2007, 82(1), 84–91.

[41] Perez-Gago, M.B. & Krochta, J.M. Denaturation time and temperature effects on solubility, tensile properties, and oxygen. Journal of Food Science. 2001, 66, 705–710.

[42] Rajam, R., Karthik, P., Parthasarathi, S., Joseph, G.S. & Anandharamakrishnan, C. Effect of whey protein-alginate wall systems on survival of microencapsulated *Lactobacillus plantarum* in simulated gastrointestinal conditions. Journal of Functional Foods. 2012, 4(4), 891–898.

[43] Rajam, R. & Anandharamakrishnan, C. Microencapsulation of *Lactobacillus plantarum* (MTCC 5422) with fructooligosaccharide as wall material by spray drying. LWT-Food Science and Technology. 2015, 60(2), 773–780.

[44] Zheng, X., Fu, N., Duan, M., Woo, M.W., Selomulya, C. & Chen, X.D. The mechanisms of the protective effects of reconstituted skim milk during convective droplet drying of lactic acid bacteria. Food Research International. 2015, 76, 478–488.

[45] Bustamante, M., Villarroel, M., Rubilar, M. & Shene, C. Lactobacillus acidophilus La-05 encapsulated by spray drying: Effect of mucilage and protein from flaxseed (*Linum usitatissimum L.*). LWT—Food Science and Technology. 2015, 62, 1162–1168.

[46] Pavlath, A. E. & Orts, W. Edible films and coatings: Why, what and how? In: Huber, K. C. & Embuscado, M. E., (Eds.), Edible Films and Coatings for Food Applications. 2009. New York: Springer; chapter 1.

[47] McHugh, T.H. Protein-lipid interactions in edible films and coatings. Nahrung. 2000, 44, 148–151.

[48] Falguera, V., Quintero, J.P., Jiménez, A., Aldemar Muñoz, J. & Ibarz, A. Edible films and coatings : Structures, active functions and trends in their use. Trends in Food Science & Technology. 2011, 22, 292–303.

[49] Rhim, J.W. & Ng, P.K.W. Natural biopolymer-based nanocomposite films for packaging applications. Critical Reviews in Food Science and Nutrition. 2007, 47(4), 411–433.

[50] Hernandez-Izquierdo, V.M. & Krochta, J.M. Thermoplastic processing of proteins for film formation—A review. Journal of Food Science. 2008, 73(2), 30–39.

[51] Tavera-Quiroz, M. J., Romano, N., Mobili, P., Pinotti, A., Gómez-Zavaglia, A., & Bertola, N. (2015). Green apple baked snacks functionalized with edible coatings of methylcellulose containing Lactobacillus plantarum. Journal of Functional Foods, 16, 164–173.

[52] Hansen, N.M.L. & Plackett, D. Sustainable films and coatings from hemicelluloses: A review. Biomacromolecules. 2008, 9(6), 1493–1505.

[53] Jiménez, A., Fabra, M.J., Talens, P. & Chiralt, A. Edible and biodegradable starch films: A review. Food and Bioprocess Technology. 2012, 5(6), 2058–2076.

[54] Elsabee, M.Z. & Abdou, E.S. Chitosan based edible films and coatings: A review. Materials Science and Engineering. 2013, 33(4), 1819–1841.

[55] Salgado, P.R., López-Caballero, M.E., Gómez-Guillén, M.C., Mauri, A.N. & Montero, M.P. Sunflower protein films incorporated with clove essential oil have potential application for the preservation of fish patties. Food Hydrocolloids. 2013, 33(1), 74–84.

[56] Rhim, J.-W. & Ng, P.K.W. Natural biopolymer-based nanocomposite films for packaging applications. Critical Reviews in Food Science and Nutrition. 2007, 47(4), 411–33.

[57] Silva-Weiss, A., Ihl, M., Sobral, P.J.A., Gomez-Guillen, M. C. & Bifani, V. Natural additives in bioactive edible films and coatings: Functionality and applications in foods. Food Engineering Reviews. 2013, 5(4), 200–216.

[58] Fito, P. Modelling of vacuum osmotic dehydration of food. Journal of Food Engineering. 1994, 22, 313–328.

[59] Fito, P. & Pastor, R. Non-diffusional mechanism occurring during vacuum osmotic dehydration (VOD). Journal of Food Engineering. 1994, 21, 513–519.

[60] Alzamora, S.M., Castro, M.A., Vidales, S.L., Nieto, A.B. & Salvatori, D. The role of tissue microstructure in the textural characteristics of minimally processed fruits. In: Minimally Processed Fruits and Vegetables. 2000. Aspen Publication: Maryland, USA, 153–171.

[61] Garg, A., Garg, S., Zaneveld, L.J.D. & Singla, A.K. Chemistry and pharmacology of the citrus bioflavonoid hesperidin. Phytotherapy Research. 2001, 15, 655–669.

[62] Betoret, E., Sentandreu, E., Betoret, N., Codoñer-Franch, P., Valls-Bellés, V. & Fito, P. Technological development and functional properties of an apple snack rich in flavonoid from mandarin juice. Innovative Food Science and Emerging Technologies. 2012, 16, 298–304.

[63] Gilles, G. Dry cured ham quality as related to lipid quality of raw material and lipid changes during processing: A review. Grasas y Aceites. 2009, 60(3), 297–307.

[64] Telahigue, K., Hajji, T., Rabeh, I. & El Cafsi, M. The changes of fatty acid composition in sun dried, oven dried and frozen hake (Merluccius merluccius) and sardinella (Sardinella aurita). African Journal of Biochemistry Research. 2013, 7(8), 158–164.

[65] Packer, L. & Ong, A.S.H. Biological oxidants and antioxidants: Molecular mechanisms and health effects. 1998. AOCS Press. Champaign: IL.

[66] Lupano, C.E. Modificaciones de componentes de los alimentos: cambios químicos y bioquímicos por procesamiento y almacenamiento. Editorial de la Universidad Nacional de La Plata, 2013.

[67] Aubourg, S. & Ugliano, M. Effect of brine pre-treatment on lipid stability of frozen horse mackerel (*Trachurus trachurus*). European Food Research & Technology. 2002, 215(2), 91–95.

[68] Baik, M.Y., Suhendro, E.L., Nawar, W.W., McClements, D.J., Decker, E.A. & Chinachoti, P. Effects of antioxidants and humidity on the oxidative stability of microencapsulated fish oil. Journal of the American Oil Chemists' Society. 2004, 81(4), 355–360.

[69] Başlar, M., Kılıçlı, M., Toker, O.S., Sağdıç, O. & Arici, M. Ultrasonic vacuum drying technique as a novel process for shortening the drying period for beef and chicken meats. Innovative Food Science & Emerging Technologies. 2014, 26, 182–190.

[70] Darvishi, H., Azadbakht, M., Rezaeiasl, A. & Farhang, A. Drying characteristics of sardine fish dried with microwave heating. Journal of the Saudi Society of Agricultural Sciences. 2013, 12(2), 121–127.

[71] Awad, T.S., Moharram, H.A., Shaltout, O.E., Asker, D. & Youssef, M.M. Applications of ultrasound in analysis, processing and quality control of food: A review. Food Research International. 2012, 48, 410–427.

[72] Babić, J., Cantalejo, M.J. & Arroquib, C. The effects of freeze-drying process parameters on Broiler chicken breast meat. LWT-Food Science and Technology. 2009, 42(8), 1325–1334.

[73] Bohm, V., Puspitasari-Nienaber, N., Ferruzzi, M.G. & Schwartz, S.J. Trolox equivalent antioxidant capacity of different geometrical isomers of α-carotene, β-carotene, lycopene and zeaxanthin. Journal of Agricultural and Food Chemistry. 2002, 50(1), 221–226.

[74] Hiranvarachat, B., Suvarnakuta, P. & Devahastin, S. Isomerisation kinetics and antioxidant activities of β-carotene in carrots undergoing different drying techniques and conditions. Food Chemistry. 2008, 107(4), 1538–1546.

[75] Chen, B.H., Peng, H.Y. & Chen, H.E. Changes of carotenoids, color, and vitamin A contents during processing of carrot juice. Journal of Agriculture and Food Chemistry. 1995, 43(7), 1912–1918.

[76] Periago, M.J. Mart 00EDnez-Valverde, I., Ros, G., Martínez, C. & L 00F3pez, G. Chemical and biological properties and nutritional value of lycopene. Anales de. Veteterinaria. (Murcia). 2001, 17, 51-66.

[77] Hedrén, E., Diaz, V. & Svanberg, U. Estimation of carotenoid accessibility from carrots determined by an in vitro digestion method. European Journal of Clinical Nutrition. 2002, 56, 425–430.

[78] Heredia, A., Peinado, I., Barrera, C. & Andrés, A. Influence of process variables on colour changes, carotenoids retention and cellular tissue alteration of cherry tomato during osmotic dehydration. Journal of Food Composition and Analysis. 2009, 22(4), 285–294.

[79] Quiñones, M., Miguel, M. & Aleixandre, A. Los polifenoles, compuestos de origen natural con efectos saludables sobre el sistema cardiovascular. Nutrición Hospitalaria. 2012, 27(1), 76–89.

[80] Wojdyło, A., Figiel, A. & Oszmiański, J. Effect of drying methods with the application of vacuum microwaves on the bioactive compounds, color, and antioxidant activity of strawberry fruits. Journal of Agriculture and Food Chemistry. 2009, 57(4), 1337–1343.

[81] López, J., Uribe, E., Vega-Gálvez, A., Miranda, M., Vergara, J., González, E. & Di Scala, K. Effect of air temperature on drying kinetics, vitamin C, antioxidant activity, total phenolic content, non-enzymatic browning and firmness of blueberries Variety O′Neil. Food Bioprocess Technology. 2010, 3, 772–777.

[82] Tamanna, N. & Mahmood, N. Food processing and Maillard reaction products: Effect on human health and nutrition. International Journal of Food Science. 2015, 2015(526762), 1–6.

[83] Lupano, C.E. Modificaciones de componentes de los alimentos: cambios químicos y bioquímicos por procesamiento y almacenamiento. ed. La Plata. Universidad Nacional de La Plata, 2013.

[84] Mujumdar, A.S. Hand book of industrial drying. 3rd ed. 2006. New York: Marcel Dekker.

[85] Beedie, M. Energy saving – a question of quality. South African Journal of Food Science Technology. 1995, 48(3), 14–16.

[86] Aghbashlo, M., Mobli, H., Rafiee, S. & Madadlou, A. A review on exergy analysis of drying processes and systems. Renewable and Sustainable Energy Reviews. 2013, 22, 1–22.

[87] Nazghelichi, T., Aghbashlo, M. & Kianmehr, M.H. Optimization of an artificial neural network topology using coupled response surface methodology and genetic algorithm for fluidized bed drying. Computers and Electronics in Agriculture. 2011, 75(1), 84–91.

[88] Singh, R.P. Energy consumption and conservation in food sterilization. Food Technology. 1977, 31, 57–60.

[89] Dincer, I. On energetic, exergetic and environmental aspects of drying systems. International Journal of Energy Research. 2002, 26, 717–727.

[90] Pandey, A.K., Tyagi, V.V., Park, S.R. & Tyagi, S.K. Comparative experimental study of solar cookers using exergy analysis. Journal of Thermal Analysis and Calorimetry. 2012, 109, 425–431.

[91] Kaushik, S.C, Siva Reddy, V. & Tyagi, S.K. Energy and exergy analyses of thermal power plants: A review. Renewable and Sustainable Energy Reviews. 2011, 15, 1857–1872.

[92] Siva Reddy, V., Kaushik, S.C. & Tyagi, S.K. Exergetic analysis of solar concentrator aided natural gas fired combined cycle power plant. Renewable Energy. 2012, 39, 114–125.

[93] Aviara, N.A., Onuoha, L.N., Falola, O.E. & Igbeka, J.C. Energy and exergy analysis of native cassava starch drying in a tray dryer. Energy. 2014, 73, 809–817.

Chapter 4

Evaporative Drying of Low-Rank Coal

Saban Pusat, Mustafa Tahir Akkoyunlu and
Hasan Hüseyin Erdem

Additional information is available at the end of the chapter

http://dx.doi.org/10.5772/63744

Abstract

Low-rank coals including the brown and the subbituminous coals are commonly known to contain high moisture content (up to 65%, wet basis), which limits their utilization around the world in spite of their low cost. Today, the most of the drying technologies are based on the evaporation of the water from the moist product. In this chapter, the most effective parameters on the evaporative coal-drying process are investigated with the data in the recent literature. The effective parameters are evaluated in three categories as follows: (1) the parameters about the drying media (the type of the media, the temperature, the pressure, the velocity and the relative humidity), (2) the coal parameters (the type of the coal and the size) and (3) the drying method.

Keywords: low-rank coal, lignite, drying, evaporative drying, coal

1. Introduction

Today, the lignite is one of the cheapest energy sources [1, 2]. The lignite reserves constitute about 45% of the total coal reserves and are distributed throughout the world [3]. The low-rank coals (LRCs) including the brown and the subbituminous coals, which are known to contain high moisture content (up to 65%, wet basis), are very important for the LRC-fired power plants, the gasification and the liquefaction [4]. The high moisture content of the LRC limits its availability in spite of its low cost [5].

The moisture in the coal causes problems in the handling, the storage, the transportation, the milling and the combustion [4, 6]. In the coal combustion, the important part of the energy is consumed to evaporate the moisture inside the coal [5–7]. The combustion of the high moisture content coal creates some problems such as the additional energy consumption for the moisture evaporation, the insufficient combustion and the additional exhaust discharge [8]. The LRC should be dried to the required moisture level to decrease the energy losses and the transportation costs, and to increase the quality of the products [9, 10]. The drying of the LRC may be divided into the evaporative drying or the non-evaporative dewatering [11]. In this study, only the evaporative drying of the LRC is considered.

The drying of the LRC decreases the problems caused by the high moisture content. In a coal-fired power plant with the coal drying, the heat lost with the flue gas, the water consumption in the cooling tower and the energy consumption in the mill decrease [12]. The efficiency of the coal-drying process for a coal-fired power plant mainly depends on the source of the drying energy. The low-quality heat source for the drying process can enhance the efficiency of a coal-fired power plant [13].

In the drying process, both the heat and the mass transfer mechanisms are active. In the evaporative drying of the coal, the heat is provided to remove the water from the coal particle. In references [5, 14, 15], it is stated that the effective parameters on the drying of the lignite are the temperature, the drying media flow rate, the sample thickness and the particle size. Many studies have been conducted on the lignite drying. In the literature, there are some attempts to review the studies about the coal drying such as references [11, 16–22].

The estimation of the exit coal moisture content of the dryer is an important research topic. However, there is not much study on this issue. The thin-layer drying models and the neural network methods were applied to estimate the drying curve [23–29]. The performance of the used models and methods seem so satisfactory.

There are various studies on the evaporative coal drying. In this study, the most effective parameters on the evaporative coal-drying process are investigated with data in recent literature open to the authors. The effective parameters are evaluated in three categories as follows: (1) the parameters about drying media, (2) the coal parameters and (3) the drying method. The effective parameters on the drying media are the type of media, the temperature, the pressure, the velocity and the relative humidity. Different coals in varying sizes are investigated in the section of parameters about coal. Finally, the drying methods used in the literature are studied. The main aims of this study are to summarize the recent studies on the LRC drying and to investigate the most effective parameters on the drying.

2. Parameters about drying media

In this section, the most effective parameters on the drying media are examined. These parameters are as follows: the type of drying media, the temperature, the pressure, the velocity and the relative humidity. All of these parameters should be defined before the design of the dryer.

2.1. Parameters about drying media

In the coal-drying literature, four drying medium (air, steam, exhaust gases and nitrogen) are used in the studies. The summary of the types of drying medium used in the coal-drying studies is presented in **Table 1**.

Drying media	References
Air	[3, 4, 6, 8, 9, 23, 26–46]
Steam	[8, 9, 25, 27, 37, 47–52, 54, 55]
Exhaust gases	[7, 54]
Nitrogen	[25, 31, 33, 53, 56, 57]

Table 1. Types of drying medium used in the literature.

The high temperature (700–900°C) air or the exhaust gases are used in the conventional evaporative dryers [18]. In the power plants, the exhaust gases can be used in the drying process, so the overall efficiency of the plant can be increased [36]. Akkoyunlu et al. [58] studied the economic upper limit of a possible dryer for the coal-fired power plants without considering the method, the conditions, the source of energy, etc.. However, in the coal drying, the air and the exhaust gases may cause some problems. The air and the exhaust gases with the high temperatures are not applicable because of the spontaneous combustion of the coal and the loss of the volatiles [11, 59].

Using superheated steam as the drying media has many advantages over the air and the exhaust gases [18, 60, 61]. The energy consumption in the air drying is more than the superheated steam drying [25]. In the superheated steam drying, the risk of oxidation and the fire are highly unlikely due to the oxygen-free atmosphere [60, 61]. Therefore, the drying temperature can be raised and the higher drying rates can be achieved [18]. The exhaust of the superheated steam drying is pure steam, and so its latent heat can be recovered by the condensation [8, 49, 53]. Moreover, using superheated steam for the coal drying with high capacities in the power plants seems more effective than the others [6].

Using nitrogen as the drying media is not applicable. However, the results of these studies can be evaluated in conjunction with the exhaust gases. The significant proportion of the exhaust gases are nitrogen.

Drying with the air and the steam are the most important topics in the coal-drying literature. The pros and cons for both are presented in many papers. In **Figure 1**, the drying rate curves for the lignite in the hot air and superheated steam are shown. For the same drying temperatures (120, 140 and 160°C), the final moisture content in the air drying is nearly zero. However, in the superheated steam drying, the final moisture content is about 0.7 kg/(kg db). The drying rate increases as the temperature increases. At the temperature of 120°C, the air drying is faster but at the temperatures of 140 and 160°C, the steam drying is faster.

Inversion temperature term is used in the comparison of the air and the steam drying. It shows the temperature point above which the drying rate in the steam is greater than that in the air. In reference [13], the inversion temperature was found in the range from 120 to 140°C.

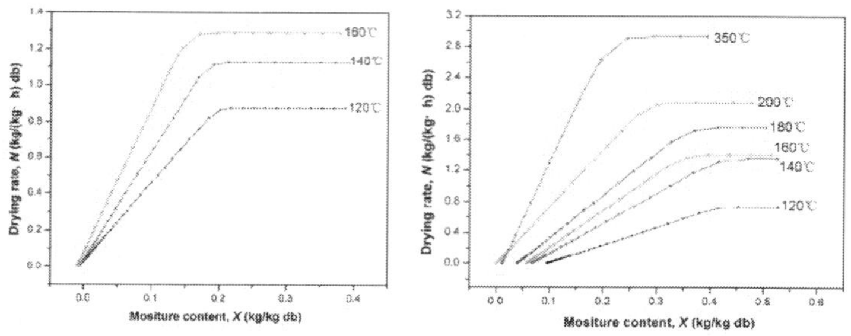

Figure 1. Drying rate curves for lignite in hot air (right) and superheated steam (left) [8].

2.2. Temperature

The drying temperature is one of the most important parameters affecting the drying rate and time. Using the high-temperature drying media requires short drying time. However, the high-temperature values are not applicable for the coal drying due to the spontaneous ignition and the loss of volatiles [59]. The drying temperature levels used in the literature are categorized in two classes (below and above the boiling temperature), and they are presented in **Table 2**.

Temperature	References
Below boiling temperature	[4, 6, 23, 28, 29, 32, 34–36, 38–41, 46, 57]
Above boiling temperature	[3, 4, 6–9, 23, 25–31, 33, 37, 38, 43–53, 55, 56]

Table 2. Drying temperature levels used in the literature.

The LRC is liable to the spontaneous combustion because of its reactive nature [62]. The high-temperature media comprising oxygen may result in combustion of the coal. Using the air or the exhaust gases (comprising uncontrolled rate of oxygen), the drying media may cause the spontaneous combustion of the coal even in the low temperatures. In some of the applications, the rate of oxygen in the exhaust gases is regulated, so the risk of the fire is controlled. However, there is still risk of the fire.

In addition, in the high temperatures, the coal losses its volatiles, which in turn decreases its calorific value [59]. Moreover, the volatiles increase the risk of the fire.

The effects of the drying temperature on the coal weight loss and the drying rate are shown in **Figure 2**. As can be seen, the higher temperature provides faster drying and short drying time.

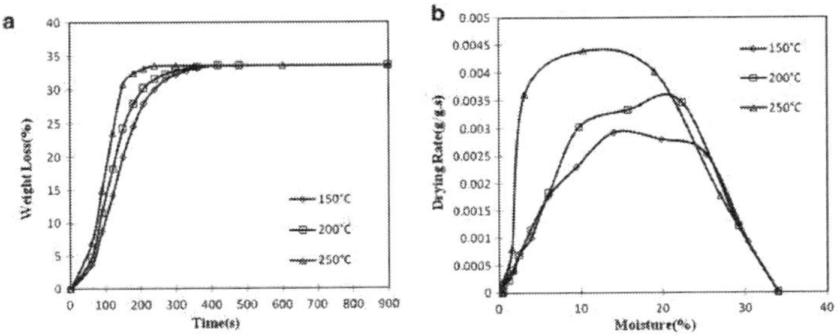

Figure 2. Change of coal weight loss with time (a) and drying rate with moisture content (b) at different temperatures [33].

2.3. Pressure

The pressure of the drying media also affects the drying of the LRC. The increase in the pressure improves the overall heat transfer coefficient [49]. However, the higher pressure values result in the higher equilibrium moisture content. The effect of the pressure can be investigated in three categories such as the atmospheric, the vacuum and the high pressure. The drying pressure levels used in the literature are presented in **Table 3**.

Pressure	References
Atmospheric	[3–6, 8, 9, 23–41, 43–47, 50–53, 55–57]
Vacuum	[47]
High	[6, 7, 47–51]

Table 3. Drying pressure levels used in the literature.

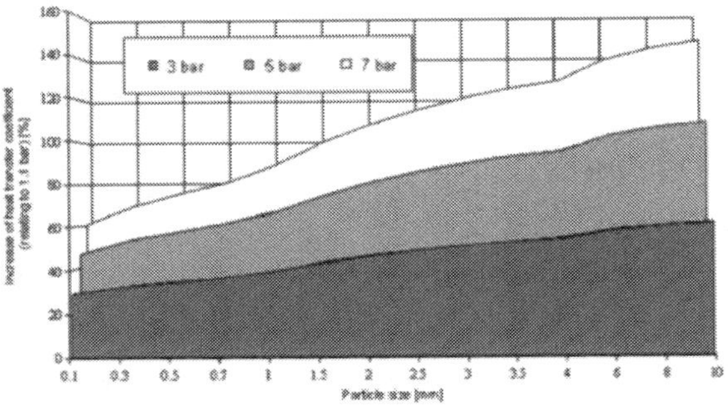

Figure 3. Effect of pressure on heat transfer coefficient [49].

The effect of the pressure on the heat transfer coefficient is shown in **Figure 3**. The percent increase in the heat transfer is calculated relating to the pressure at 1.1 bar. The pressure seems significantly effective according to reference [49]. Moreover, according to reference [63], the higher pressure values result in the faster drying. However, according to reference [48], the pressure does not affect the drying rate. The effect of the pressure on the drying should be presented clearly.

2.4. Velocity

The velocity of the drying media is effective on the LRC drying. In the literature, the different velocity values are studied. In the fluid bed coal-drying studies, the fluidization velocity is also studied. For the case of the fluid bed drying, the level of the drying media's velocity according to the minimum fluidization velocity is very important.

The effect of the drying media is investigated in reference [39] (**Figure 4**). The higher velocity value provides the faster drying rate. The velocity does not affect the drying rate significantly in the last part of the drying. For the fluidized bed dryers, the higher fluid velocities increase the heat transfer rate and the solid mixing [33].

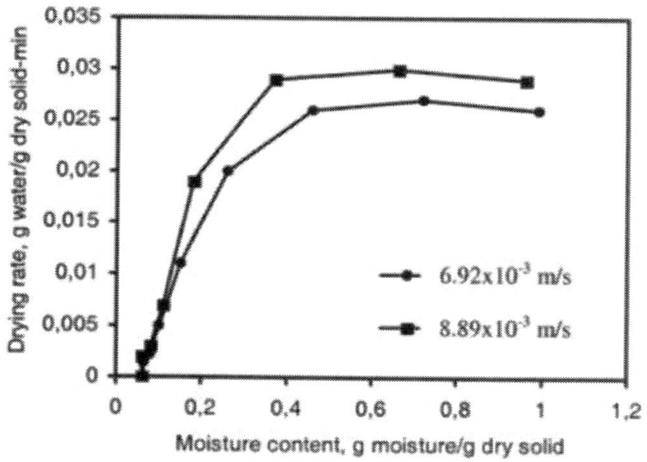

Figure 4. Effect of velocity on coal drying [39].

2.5. Relative humidity

The relative humidity of the drying media affects the drying of the LRC. As can be seen from **Figure 5**, the lower relative humidity means the higher drying rate. At the surface of the coal particles, the evaporation rate is dependent on the water vapour pressure difference between the coal surface and the drying media. The water vapour pressure of the drying media increases with the increase in humidity, and thus, the drying rate decreases with the increase in humidity. In addition, the equilibrium moisture content of the coal particles increases with the increase in the relative humidity.

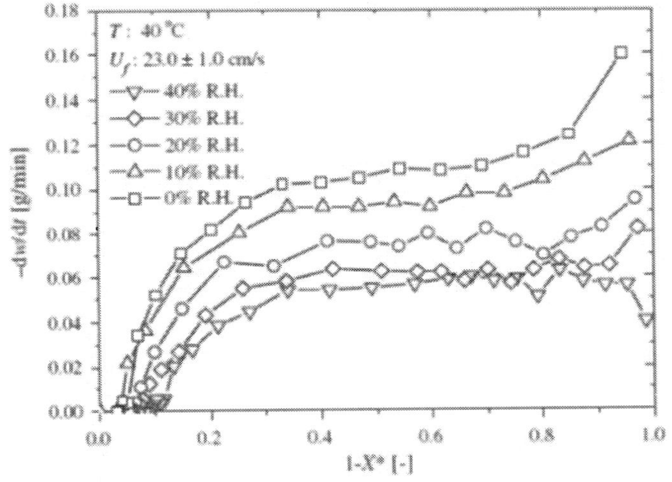

Figure 5. Effect of relative humidity on coal drying [41].

3. Parameters about coal

The characteristics and the particle size of the coal have an important effect on the drying. All types of the coals have different characteristics such as the initial moisture content, the porosity, the equilibrium moisture content, the volatile matter, the grindability, the ash content and the heating value. The effect of the moisture content on the coal heating value for different coal types is shown in **Figure 6**.

The moisture in the coal can be categorized in three groups: the surface moisture (the free water), the physically bound moisture and the chemically bound moisture [32, 64]. The heat is provided to the coal particle, in the evaporative drying, for heating the particle, for evaporating the water, and for overcoming the binding forces (both the physical and the chemical) between the coal and the water [32, 49]. The surface water is easily removed by the evaporation but the other types of the moisture require more energy to be removed. As can be seen from **Figure 7**, the heat of the desorption of the water from the Yallourn brown coal increases with the decrease in the moisture content after a critical moisture value, which shows the end of the surface water and the start of the domination of the internal mass transfer mechanisms.

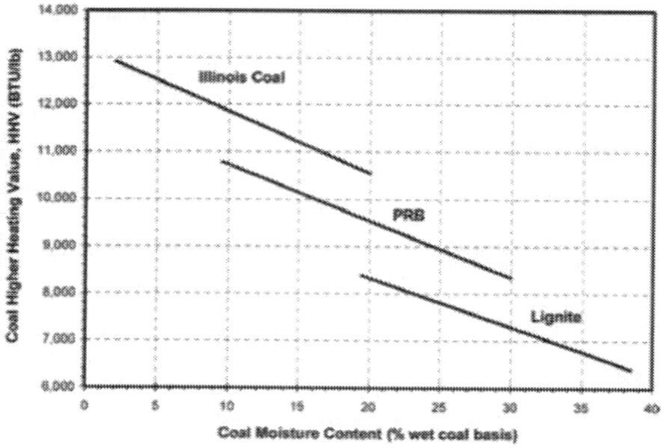

Figure 6. Coal heating value as a function of coal moisture content [32].

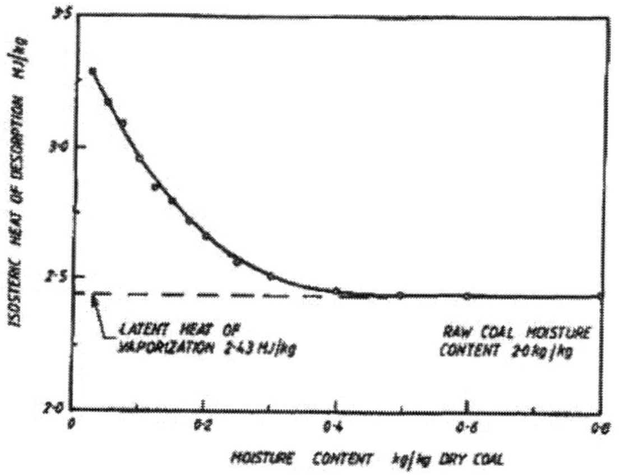

Figure 7. Heat of desorption of water as a function of moisture content [65].

It is important to understand the types of the water in the coal to be effectively removed. For different coal types, the binding forces change, and the binding enthalpy increases with the decreasing moisture content (**Figure 8**) [66–68].

The higher part of the water in the lignite is in the pores [69]. Therefore, the number, the size, the distribution and the shape of the pores in the LRC have important effects on the drying. The water in the smaller pores means difficult to remove. The importance of the effects of the coal parameters on the evaporative drying clarifies that all the types of the coal should be studied separately to obtain the drying characteristics.

http://dx.doi.org/10.5772/63744

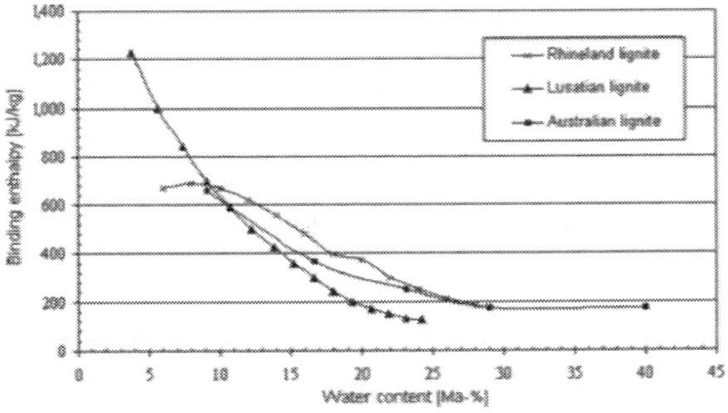

Figure 8. Change of binding enthalpy with coal and water content [49].

3.1. Type of coal

In the literature, there are many studies ([5, 7, 30, 32, 35, 36, 45, 50, 51, 57, 70], etc.), which investigated the effects of the coal type on the drying. In **Figure 9**, the drying curves of the North Dakota lignite and the subbituminous coal from the Powder River Basin (PRB) are shown. Different types of coals show different drying characteristics.

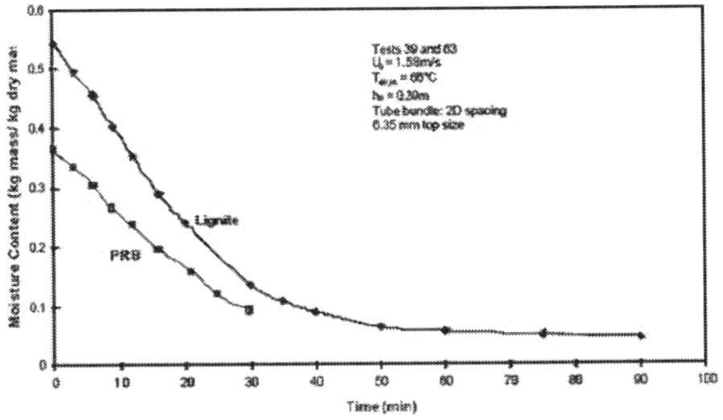

Figure 9. Comparison of drying curves for two different coals [35].

3.2. Particle size

The particle size is highly important in the drying process. In addition, the particle size is very important parameter in the fluidization of the fluidized bed. Moreover, the size of the lignite particles has an important effect on the heat transfer coefficient inside the superheated steam fluidized bed dryer [49]. The sizes of the coal particles used in the literature are presented in **Table 4**.

Particle size (mm)	References
<2	[5–8, 24, 25, 27, 33, 34, 39–41, 47, 53, 57]
<5	[3, 4, 9, 26, 31, 32, 35–37, 44, 49–52, 56]
>5	[23, 26, 28–30, 35–37, 43, 45, 46, 48, 49, 55]

Table 4. Coal particle sizes studied in the literature.

The effect of the particle size on the drying rate is presented in **Figure 10**. The drying rate increases as the coal particle size decreases. The smaller particle fractions have larger surface area, and thus, they dry faster [71]. In addition, the moisture transport distance inside the particle decreases as the particle size decreases.

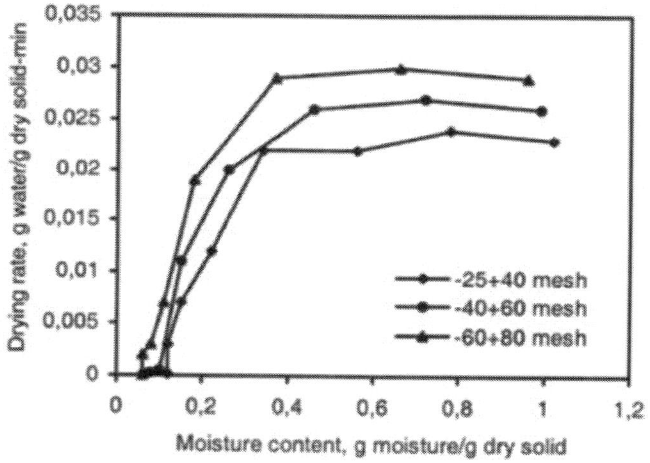

Figure 10. Effect of particle size on drying rate [39].

4. Drying method

In the literature, many different coal-drying technologies are seen. However, the common conventional drying systems are the fluidized bed dryer, the rotary dryer, the shaft dryer, the pneumatic dryer, fixed bed, etc. [72]. The experimental drying methods used in the literature are presented in **Table 5**.

Some of the experimental drying methods (the TGA-Thermal Gravimetric Analysis, the oven and the others) used in the literature are just for investigating, analysis and modelling. Therefore, they are not examined for the applicability of the methods. The results of these studies are important to understand the drying characteristics of the coals, to evaluate the effective parameters on the drying and to be able to model the drying of the LRC in a convenient drying technology.

Drying method	References
Fluidized bed	[3, 9, 25–27, 30–33, 35, 36, 38–41, 47–49, 53]
Microwave	[5, 24, 25, 53]
Moving bed	[43]
Flash	[7]
TGA	[34, 44, 56, 57]
Oven	[4, 6, 37, 50–52]
Fixed bed	[23, 28, 29, 46]
Others	[8, 44, 55]

Table 5. Experimental drying methods used in the literature.

The fluidized bed method is extensively used in the drying of the wet particulate and the granular materials [39, 72]. It has many advantages such as the better gas-drying medium contact, the high thermal efficiency and the drying rates [15]. However, it has some disadvantages such as the high-pressure drops and the non-uniform moisture in the output products [73].

The low-temperature fluidized bed drying method is developed in the United States [32, 74]. The low-grade waste heat is used in this process. This method decreases the risk of the oxidation and the fire due to the low-temperature air. The in-bed heat exchangers are used to increase the temperature of the air and its moisture carrying capacity. However, there is still a risk of the spontaneous combustion in the low-temperature air drying.

The superheated steam fluidized bed-drying technology is a promising one for the coal drying, especially for the high capacities such as the coal-fired power plants. For the power plants, the necessary steam for the drying process can be supplied from the turbine. The in-bed heat exchangers are also used in this method. The heat is supplied to the exchanger tubes by the steam in the lignite drying process [49]. The generated steam can be used in the process by increasing its temperature by the vapour compressor [49]. In addition, the generated condensate in the in-bed heat exchangers can be used for preheating [49]. Using the steam as the drying and the heating medium may increase the efficiency of the process considerably.

The microwave drying is used in a few coal-drying studies [5, 53]. It has some advantages such as the higher heating rates compared to the conventional heating and the more uniform heat supply [5, 60]. The microwave drying directly uses the electricity as the energy source, so it seems so expensive for the LRC drying. However, it can be used by integrating with a conventional drying system due to the advantages of the microwave drying in removing water inside the coal, which is difficult to evaporate with the other drying technologies (**Figure 11**) [75].

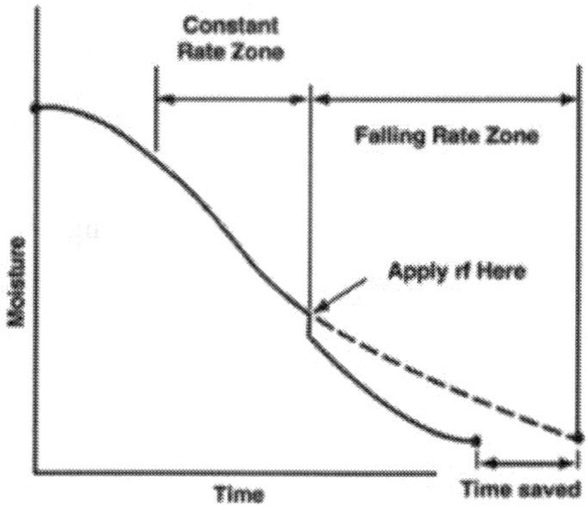

Figure 11. Effect of microwave drying on normal drying curve [75].

The microwave power level has effects on the drying with the microwave. The weight loss increases with the increasing microwave power (**Figure 12**). In addition, the coal type affects the weight loss with the microwave drying.

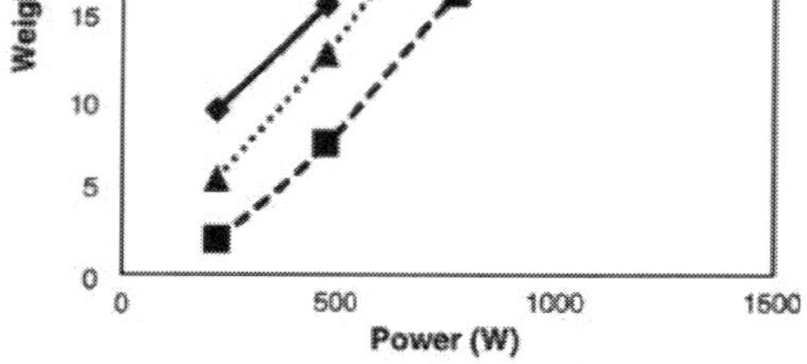

Figure 12. Effect of microwave power level on coal weight loss for three different coal types [5].

The flash drying is one of the most widely used technologies in the drying, and it is also known as the pneumatic drying [72]. As the drying medium, the steam, the air and the exhaust gases can be used. It requires the high drying medium velocities to transport the particles. The particle size range for the flash drying is usually 0.01–0.5 mm [72]. It is not applicable for the larger particle sizes.

The packed moving bed dryer was developed for the large capacities and it uses the 150°C exhaust gases with the controlled oxygen content [76]. Using of the exhaust gases at this temperature levels provides the waste heat recovery potential. In addition, the controlled oxygen content decreases the spontaneous combustion risk. This methodology seems applicable to the power plants with the high capacities and the heat recovery systems.

The fixed-bed drying technology was used to dry the coarse lignite particles [46]. However, there is not much study on the fixed-bed drying. This methodology is a promising one for the drying of lignite particles greater than 10 mm.

5. Conclusions

The coal is one of the most important energy sources in the world. The drying of the LRC is very essential to utilize it efficiently. Because the coal has a reactive and a combustible nature, the drying technology and the drying media should be determined carefully. Moreover, drying is an energy-intensive process; thus, the energy source for the drying process should be determined with care. There is not only one correct method to dry a product. There are numerous methods in the drying literature. Therefore, every drying process should be studied separately.

The drying of the LRCs is still a contemporary topic. There are many studies on the LRC drying, but the current studies are not enough. According to the authors, some further steps should be taken as stated below:

- There is not any detailed study, which examines the effect of porosity on the drying characteristics of the LRCs and links up the drying characteristics, the coal type and the porosity.
- The single-particle drying characteristics of different coals should be studied in detail, and all the effective parameters (particularly the pressure) on the drying should be presented clearly.
- One of the most important areas for the LRCs is the power plants. More elaborative studies should be conducted over the use of the LRC in the power plants such as:

  ○ All the stages (from the mine to the boiler) of the coal combustion in the power plant should be presented clearly, and the problems caused by the moisture content should be presented.

  ○ All the possible drying technologies should be presented.

  ○ All the possible energy sources should be presented, and especially the waste heat recovery sources should be determined.

  ○ For the technologies using the superheated steam, the possibilities of using the process steam should be evaluated, and the optimization studies should be conducted.

- The innovative coal drying systems should be created, and the hybrid and integrated coal-drying systems should be examined.
- There are many mathematical, numerical and theoretical models for the drying of the moist solids. The simple models should be developed for the coal drying.

Acknowledgements

The authors thank to Prof. İsmail Ekmekçi, Prof. İsmail Teke and Prof. Bahri Şahin for their valuable comments and contributions.

Author details

Saban Pusat*, Mustafa Tahir Akkoyunlu and Hasan Hüseyin Erdem

*Address all correspondence to: spusat@yildiz.edu.tr

Department of Mechanical Engineering, Yildiz Technical University, Besiktas, Istanbul, Turkey

References

[1] Kavouridis K. Lignite industry in Greece within a world context: mining, energy supply and environment. Energy Policy. 2008;36:1257–1272.

[2] Kozłowski Z. Present situation and prospects for lignite in the Polish power generation industry. Applied Energy. 2003;74:323–329.

[3] Jeon D, Kang T, Kim H, Lee S, Kim S. Investigation of drying characteristics of low rank coal of bubbling fluidization through experiment using lab scale. Science China Technological Sciences. 2011;54(7):1680–1683.

[4] Xianchun L, Hui S, Qi W, Chatphol M, Terry W, Jianglong Y. Experimental study on drying and moisture re-adsorption kinetics of an Indonesian low rank coal. Journal of Environmental Sciences. 2009;21(1):127–130.

[5] Tahmasebi A, Yu J, Li X, Meesri C. Experimental study on microwave drying of Chinese and Indonesian low-rank coals. Fuel Processing Technology. 2011;92:1821–1829.

[6] Pakowski Z, Adamski R, Kokocinska M, Kwapisz S. Generalized desorption equilibrium equation of lignite in a wide temperature and moisture content range. Fuel. 2011;90:3330–3335.

[7] Ross D, Doguparthy S, Huynh D, Mcintosh M. Pressurised flash drying of Yallourn lignite. Fuel. 2005;84:47–52.

[8] Shi Y, Li J, Li X, Wu J, Wu M, Li S, Wang H, Zhao G, Yin F. Experimental study on super-heated steam drying of lignite. Advanced Materials Research. 2012;347–353:3077–3082.

[9] Chen Z, Agarwal PK, Agnew JB. Steam drying of coal. Part 2. Modeling the operation of a fluidized bed drying unit. Fuel. 2001;80(2):209–223.

[10] Kakaras E, Ahladas P, Syrmopoulos S. Computer simulation studies for the integration of an external dryer into a Greek lignite-fired power plant. Fuel. 2002;81(5):583-593.

[11] Allardice DJ, Chaffee AL, Jackson WR, Marshall M. Chapter 3—Water in brown coal and its removal. In: Li C-Z, editor. Advances in the Science of Victorian Brown Coal. 2004. Netherlands: Elsevier, pp. 5–133.

[12] Levy EK, Sarunac N, Bilirgen H, Caram H. Use of coal drying to reduce water consumed in pulverized coal power plants. Final Report 2006. Energy Research Center Lehigh University 117 ATLSS Drive Bethlehem, PA 18015, DOE Report DE-FC26-03NT41729.

[13] Trent H, Andrew H, Barry H. Process integration analysis of a brown coal-fired power station with CO_2 capture and storage and lignite drying. Energy Procedia. 2009;1(1): 3817–3825.

[14] Kannan CS, Thomas PP, Varma YBG. Drying of solids in fluidized beds. Industrial and Engineering Chemistry Research. 1995;34(9):3068–3077.

[15] Diamond NC, Magee TRA, McKay G. The effect of temperature and particle size on the fluid bed drying of Northern Ireland lignite. Fuel. 1990;69(2):189–193.

[16] Willson WG, Walsh D, (Bill) Irwinc W. Overview of low-rank coal (LRC) drying. Coal Preparation. 1997;18(1–2):1–15.

[17] Katalambula H, Gupta R. Low-grade coals: a review of some prospective upgrading technologies. Energy & Fuels. 2009;23:3392–3405.

[18] Karthikeyan M, Zhonghu W, Mujumdar AS. Low-rank coal drying technologies—current status and new developments. Drying Technology. 2009;27(3):403–415.

[19] Yang X, Zhao Y, Luo Z, Chen Z, Duan C, Song S. Brown coal drying processes—a review. In: 2011 International Conference on Materials for Renewable Energy & Environment (ICMREE); IEEE; Shanghai; 2011. pp. 377–381.

[20] Osman H, Jangam SV, Lease JD, Mujumdar AS. Drying of low-rank coal (LRC)—a review of recent patents and innovations. Minerals, Metals and Materials Technology Centre (M3TC); Singapore; 2011.

[21] Jangam SV, Karthikeyan M, Mujumdar AS. A critical assessment of industrial coal drying technologies: role of energy, emissions, risk and sustainability. Drying Technology. 2011;29(4):395–407.

[22] Yu J, Tahmasebi A, Han Y, Yin F, Li X. A review on water in low rank coals: the existence, interaction with coal structure and effects on coal utilization. Fuel Processing Technology. 2013;106:9–20.

[23] Akkoyunlu MT, Akkoyunlu MC, Pusat S, Özkan C. Prediction of accurate values for outliers in coal drying experiments. Arabian Journal for Science and Engineering. 2015;40(9):2721–2727.

[24] Pickles CA, Gao F, Kelebek S. Microwave drying of a low-rank sub-bituminous coal. Minerals Engineering. 2014;62:31–42.

[25] Tahmasebi A, Yu J, Han Y, Zhao H, Bhattacharya S. A kinetic study of microwave and fluidized-bed drying of a Chinese lignite. Chemical Engineering Research and Design. 2014;92(1):54–65.

[26] Zhao P, Zhao Y, Luo Z, Chen Z, Duan C, Song S. Effect of operating conditions on drying of Chinese lignite in a vibration fluidized bed. Fuel Processing Technology. 2014;128:257–264.

[27] Stokie D, Woo MW, Bhattacharya S. Comparison of superheated steam and air fluidized-bed drying characteristics of Victorian brown coals. Energy and Fuels. 2013;27(11):6598–6606.

[28] Pusat S, Akkoyunlu MT, Erdem H, Dağdaş A. Drying kinetics of coarse lignite particles in a fixed bed. Fuel Processing Technology. 2015;130:208–213.

[29] Pusat S, Akkoyunlu MT, Pekel E, Akkoyunlu MC, Özkan C, Kara SS. Estimation of coal moisture content in convective drying process using ANFIS. Fuel Processing Technology. 2016;147:12–17.

[30] Komatina M, Manovic V, Saljnikov A. A model of coal particle drying in fluidized bed combustion reactor. Energy Sources, Part A: Recovery, Utilization, and Environmental Effects. 2007;29:239–250.

[31] Agarwal PK, Genetti WE, Lee YY. Drying and devolatilization of Mississippi lignite in a fluidized bed. Abstracts of Papers of the American Chemical Society. 1984:187;28.

[32] Sarunac N, Levy EK, Ness M, Bullinger CW, Mathews JP, Halleck PM. A novel fluidized bed drying and density segregation process for upgrading low-rank coals. International Journal of Coal Preparation and Utilization. 2009;29(6): 317–332.

[33] Tahmasebi A, Yu J, Han Y, Li X. A study of chemical structure changes of Chinese lignite during fluidized-bed drying in nitrogen and air. Fuel Processing Technology. 2012;101:85–93.

[34] Bueno JL, Iglesias O, Garcia A. Drying of particulate solids: determination of the characteristic curve of brown coal. Drying Technology. 1993;11(3):555–570.

[35] Levy EK, Caram HS, Yao Z, Wei Z, Sarunac N. Kinetics of coal drying in bubbling fluidized beds. In: Proceedings Fifth World Congress on Particle Technology; Orlando, Florida; 2006. pp. 1–6.

[36] Wang W-C. Laboratory investigation of drying process of Illinois coals. Powder Technology. 2012;225:72–85.

[37] Chen Z, Wu W, Agarwal PK. Steam-drying of coal. Part 1. Modeling the behavior of a single particle. Fuel. 2000;79(8):961–973.

[38] Park J, Shun D, Bae D-H, Lee S, Seo JH, Park JH. The effect of gas temperature and velocity on coal drying in fluidized bed dryer. In: The 13th International Conference on Fluidization—New Paradigm in Fluidization Engineering; ECI; Korea; 2010. pp. 1–8.

[39] Çalban T, Erşahan H. Drying of a Turkish lignite in a batch fluidized bed. Energy Sources. 2003;25:1129–1135.

[40] Çalban T. The effects of bed height and initial moisture concentration on drying lignite in a batch fluidized bed. Energy Sources, Part A: Recovery, Utilization, and Environmental Effects. 2006;28(5):479–485.

[41] Kim H-S, Matsushita Y, Oomori M, Harada T, Miyawaki J, Yoon S-H, Mochida I. Fluidized bed drying of Loy Yang brown coal with variation of temperature, relative humidity, fluidization velocity and formulation of its drying rate. Fuel. 2013;105:415–424.

[42] Stakić M, Tsotsas E. Modeling and numerical analysis of an atypical convective coal drying process. Drying Technology. 2004;22(10):2351–2373.

[43] Zhang K, You C. Numerical simulation of lignite drying in a packed moving bed dryer. Fuel Processing Technology. 2012;110:122–132.

[44] Kang T-J, Namkung H, Xu L-H, Lee S, Kim S, Kwon H-B, Kim H-T. The drying kinetics of Indonesian low rank coal (IBC) using a lab scale fixed-bed reactor and thermobalance to apply catalytic gasification process. Renewable Energy. 2013;54:138–143.

[45] Zhang K, You C. Experimental and numerical investigation of convective drying of single coarse lignite particles. Energy & Fuels. 2010;24:6428–6436.

[46] Pusat S, Akkoyunlu MT, Erdem HH, Teke I. Effects of bed height and particle size on drying of a Turkish lignite. International Journal of Coal Preparation and Utilization. 2015;35(4):196–205.

[47] Potter OE, Beeby CJ, Fernando WJN, Ho P. Drying brown coal in steam-heated, steam-fluidized beds. Drying Technology. 1983;2(2):219–234.

[48] Looi AY, Golonka K, Rhodes M. Drying kinetics of single porous particles in superheated steam under pressure. Chemical Engineering Journal. 2002;87:329–338.

[49] Hoehne O, Lechner S, Schreiber M, Krautz HJ. Drying of lignite in a pressurized steam fluidized bed—theory and experiments. Drying Technology. 2010;28:5–19.

[50] Bongers GD, Jackson WR, Woskoboenko F. Pressurised steam drying of Australian low-rank coals: Part 1. Equilibrium moisture contents. Fuel Processing Technology. 1998;57:41–54.

[51] Bongers GD, Jackson WR, Woskoboenko F. Pressurised steam drying of Australian low-rank coals: Part 2. Shrinkage and physical properties of steam dried coals, preparation of dried coals with very high porosity. Fuel Processing Technology. 2000;64:13–23.

[52] Suwono A, Hamdani U. Upgrading the Indonesian's low rank coal by superheated steam drying with tar coating process and its application for preparation of CWM. Coal Preparation. 1999;21(1):149–159.

[53] Tahmasebi A, Yu J, Han Y, Zhao H, Bhattacharya S. A kinetic study of microwave and fluidized-bed drying of a Chinese lignite. Chemical Engineering Research and Design. 2014;92(1):54–65.

[54] Liu M, Yan J, Chong D, Liu J, Wang J. Thermodynamic analysis of pre-drying methods for pre-dried lignite-fired power plant. Energy. 2013;49:107–118.

[55] Pakowski Z, Adamski R, Kwapisz S. Effective diffusivity of moisture in low rank coal during superheated steam drying at atmospheric pressure. Chemical and Process Engineering. 2012;33(1):43–51.

[56] Li YH, Skinner JL. Development and validation of a process simulator for drying subbituminous coal. Chemical Engineering Communications. 1986;49(1–3):99–118.

[57] Vorres KS. Effect of temperature, sample size, and gas flow rate on drying on Beulah-zap lignite and Wyodak subbituminous coal. Energy and Fuels. 1994;8(2):320–323.

[58] Akkoyunlu MT, Erdem HH, Pusat S. Determination of economic upper limit of drying process in coal-fired power plants. Drying Technology. 2016;34(4):420–427.

[59] Jangam SV, Karthikeyan M, Mujumdar AS. A critical assessment of industrial coal drying technologies: role of energy, emissions, risk and sustainability. Drying Technology. 2011;29(4):395–407.

[60] Wang ZH, Chen G. Theoretical study of fluidized-bed drying with microwave heating. Industrial and Engineering Chemistry Research. 2000;39(3):775–782.

[61] Suvarnakuta P, Devahastin S, Soponronnarit S, Mujumdar AS. Drying kinetics and inversion temperature in a low-pressure superheated steam-drying system. Industrial and Engineering Chemistry Research. 2005;44(6):1934–1941.

[62] Karthikeyan M. Minimization of moisture readsorption in dried coal samples. Drying Technology. 2008;26(7):948–955.

[63] Olufemi BA, Udefiagbon IF. Modelling the drying of porous coal particles in superheated steam. Chemical and Biochemical Engineering Quarterly. 2010;24(1):29–34.

[64] Effenberg D. Theoretische und Praktische Untersuchungen zur Trocknung von Braunkohle [thesis]. Dresden: TU Dresden; 1989.

[65] Allardice DJ, Evans DG. The brown-coal/water system: Part 2. Water sorption isotherms on bed-moist Yalloum brown coal. Fuel. 1971;50(3):236–253.

[66] Schafer HG, Opdenwinkel H. Determination of bond enthalpy from desorption isotherms of a Rhenish brown coal in the range of high temperature. Chemiker-Zeitung. 1985;109(5):171.

[67] Wahl T, Franke B. Zum waermeverbrauch bei der braunkohletrocknung. Braunkohle. 1990 ;7:30–34.

[68] Allardice DJ, Evans DG. The brown-coal/water system: Part 1. The effect of temperature on the evolution of water from brown coal. Fuel. 1971;50(2):201–210.

[69] Karthikeyan M, Mujumdar AS. Factors affecting quality of dried low rank coals. M3TC Technical Report – Coal Drying TN-08-01; Singapore; 2007.

[70] Zhang K, You C. Experimental and numerical investigation of lignite particle drying in a fixed bed. Energy and Fuels. 2011;25(9):4014–4023.

[71] Wang Z, Chen G. Heat and mass transfer in batch fluidized-bed drying of porous particles. Chemical Engineering Science. 2000;55(10):1857–1869.

[72] Mujumdar AS. Handbook of Industrial Drying. 3rd ed. Boca Raton, FL: CRC Press; 2006.

[73] Devahastin S. Mujumdar's Practical Guide to Industrial Drying: Principles, Equipment and New Developments. Montreal: Exergex Corporation; 2000.

[74] Bullinger CW, Sarunac N. Lignite Fuel Enhancement—Final Technical Report: Phase 1. 2006. USA: U.S. Department of Energy.

[75] Schiffmann RF. Microwave and dielectric drying. In: Mujumdar AS, editor. Handbook of Industrial Drying. 3rd ed. Boca Raton, FL: CRC Press; 2006.

[76] Pang S, Xu Q. Drying of woody biomass for bioenergy using packed moving bed dryer: mathematical modeling and optimization. Drying Technology. 2010;28:702–709.

Chapter 5

Spray Drying of Xoconostle Juice: Interaction of Microstructure, Function, and Drying Operation Conditions

Jaime Jiménez-Guzmán, Diana E. Leyva-Daniel,
Brenda H. Camacho-Díaz,
Antonio R. Jimenéz-Aparicio and
Liliana Alamilla-Beltrán

Additional information is available at the end of the chapter

http://dx.doi.org/10.5772/63723

Abstract

The xoconostle fruit (*Opuntia matudae*) is rich in polysaccharides, soluble fiber, simple phenols, betalains, and ascorbic acid. However, its consumption is limited due to its high acidity. Spray drying could be a technological option to strengthen the sustainability of xoconostle giving a re-valorization as a possible natural additive for the food industry. The food powders have to be designed considering aspects related to the effect of processing conditions on final quality properties; in this case, the effect of different drying air temperatures was evaluated on moisture content, water activity (Aw), glass transition temperature, microstructure, antioxidant activity, phenolic, and betalain compounds. For all cases, the drying air temperature had a positive effect on physical stability, at low levels of water activity and moisture content, and glass transition temperature (T_g) was increased. The biological functionality (assessed through phenolics, betalain compounds, and antioxidant activity) was also kept constant for all processing conditions investigated. However, the most evident changes were observed at microscopic scale analyzed through morphometric parameters.

Keywords: spray drying, food powders, xoconostle juice, microstructure, betalains, phenolics content, antioxidant activity

1. Introduction

In food, pharmaceutical, and biotechnology industries, spray drying is the most used drying method. In this, a material in liquid state is fed by atomization in the form of fine droplets into a drying chamber, to obtain a solid product (powder). The outlet temperature of the dried product is between the wet bulb temperature and the outlet drying air temperature, remaining below 100°C [1]. Some factors that have shown effect on the final quality characteristics of food powders obtained by spray drying are the type of atomizer (two-fluid nozzle, sonic nozzle, spray nozzle, or rotatory disc), the physical properties of fed materials, heat and mass transfer phenomena, the average size of the atomized drops and their distribution, particle trajectory, size, and speed of particles [2, 3].

The food powders with high sugar content can be classified into two groups: sticky and non-sticky. The non-sticky products such as milk products and egg powders are obtained by spray drying in simple operating conditions and are characterized by free-flowing powders [4]. The powders obtained from fruit juices, honey, and lactose are examples of sticky products. These products are difficult to dry in a spray dryer due to their high sugar content (sucrose, glucose, lactose, and fructose), organic acids (citric, malic, and tartaric acid) [5], and hygroscopic characteristics. To improve the efficiency of drying process of sticky materials, agents of high molecular weight such as maltodextrins are used because of their high solubility, low viscosity, and high glass transition temperature [4, 6].

2. Drying

Drying is defined as the elimination of water in relatively small quantities of certain materials, under controlled conditions. Drying or dehydration of biological materials, in particular food products, is used as a conservation method [7]. The microorganisms that cause food decomposition cannot grow and multiply in the absence of water, and many enzymes that cause biochemical changes in the food and other biological materials may be inactive [8]. Microorganisms are inactive when the moisture content is below 10%. However, it is usually necessary to reduce the moisture content below 5% of weight in food to preserve its flavor and nutritional value [9].

3. Spray drying

Spray drying, as unitary operation, is used to transform liquid into solid particles (powder), eliminating moisture. In food industry, spray drying is widely used as an encapsulation method for food ingredients such as flavors, vitamins, minerals, dyes, waxes, and oils, in order to protect them from environmental stress and prolong the storage stability [10, 11]. Spray drying is commonly used in the food industry since, in comparison with other encapsulation techniques, its cost of production is low, has high availability of equipment, and also produces particles of good quality [12]. The fed materials (solution, emulsion, suspension) are atomized within the drying chamber in the form of very small drops (between 20 and 250 μm). This increases the total contact surface, improving the evaporation of the solvent (usually water).

When the product is encapsulated by spray drying, the nucleus exists as microparticles or droplets distributed within the dry solid capsule. According to the size of the particle, the encapsulation is classified as macro (>5000 μm), micro (0.2–5000 μm), and nano (<0.2 μm). Some authors point out that spray drying can be considered a sustainable process for the consolidation of nanoparticles within a spherical particle of micron size, but with properties in nano scale [13, 14].

Spray-drying equipment components are heating system and circulation of air, atomizer (nozzle or rotary disk), and powder recovery system. The most important stage of spray drying is atomization, because this affects the size, distribution of the droplets and, consequently, the final particle size. The most used atomizers are the rotary disc and pressure nozzle. Two-fluid pneumatic nozzles are used in special applications or in low-capacity dryers in pilot plants. Surface tension, viscosity, and density are the main characteristics of liquid that influence atomization [15]. Processing parameters such as configuration of dryer and operating conditions exert a significant influence on the properties of the final product [16, 17].

3.1. Drying stages

The main stages of a dryer spray are atomization of the fed product, contact air-drop, evaporation of moisture from the product, and recovery of the dry product [15].

a. Atomizing—this stage generates fine drops, increasing the relationship surface-mass, and it is the key parameter to determine the size of the particle. In all cases, the atomization occurs when the magnitude of the disruptive force exceeds the size of the surface tension [15].

b. Contact of drying air droplet—at this stage, the atomized droplets fall into a hot air flow inside the drying chamber. The contact can be on different arrays of liquid flow with the drying air flow (co-current, countercurrent, or mixed flow) [18].

c. Evaporation—the process of elimination of solvent contained in the droplet is divided into two phases. In the first stage, the droplet is heated when contacting with hot air, achieving a value closer to the wet bulb temperature, which corresponds to the drying air condition; in the second stage, a shell is formed, and diffusional process of water and soluble solids are detected [19].

d. Solids recovery—in the last stage, the powder is collected in a cyclone, which consists of a cylinder with a tangential opening through which the gas flows with particles generated in the dryer [20].

During evaporation stage, in the first step, the drying rate is almost constant, at which the surface of the drop remains saturated by the migration of water from the inside to the outside of the particle, and at the same time the solutes are concentrated, and the droplet diameter can be reduced. On the contrary, the drying air temperature decreases, and the particle temperature is increased [21]. In the second step, two regions of the particle are distinguished: a dried shell and a wet core. Drying of the particle speed is controlled by the moisture diffusion from the center to the surface of the shell. As a result of drying, the thickness of the crust increases, and its temperature rises to the set by the end of the process. Once the moisture content drops to a minimum value, the product is considered as a dry particle [22, 23].

3.2. Spray drying of high-sugar products

Food products dried by this process can be classified into sticky and non-sticky products. This categorization is relative, since some non-sticky products behave in an opposite manner, depending on the process conditions and hygroscopicity [24].

The non-sticky products can be treated through a simple dryer, and the powders obtained have a low hygroscopicity and can be classified as free-flow powders. The non-sticky products are dairy powders, micro-encapsulated powders, and powdered white egg. These powders can be dried in single operating conditions and are characterized by dust that flows free [5, 25].

In sticky products such as juicy fruits and vegetables, the spray drying treatment is usually difficult [24]. Sticky products hinder spray drying in normal conditions, such as powders obtained from the juice of fruits and vegetables, honey powder, and lactose powder. This is mainly due to the high content of sugars (sucrose, glucose, lactose, and fructose) and organic acids (citric, malic, and tartaric acid). During drying a product with a high content of sugar, the viscosity of the drops increases until it reaches a critical value (10^7 Pa s), making it a rubbery product prior to obtaining an amorphous structure. This kind of gumminess is considered as a sticky structure that is linked to the water activity (Aw) and temperature, which is between the glass transition temperature and the temperature at which the product is sticky (10–30°C) [26]. When the surface of the droplet reaches a sticky state, it raises the coalition with any area of the surface of another particle. This accumulation or agglomeration depends on the speed, strength, angle, and contact time. The result is considered negative for the case in which the particles adhere to the internal walls of the dryer chamber, causing a loss of the product. However, there are positive benefits when the particles are linked among them, contributing to the drying and changing its structure, forming clumps [27]. The moisture content in sticky products such as juice powder fruit must be between 2 and 4% [25]. Controlling the moisture content is very important, since it has a direct effect on the powder quality.

4. General characteristics of powders

Spray drying of liquid foods or biological materials produces amorphous particles mainly due to the rapid evaporation and the short time for crystallization [28]. Since morphological features are created by the loss of moisture, the rigidity of the surface of amorphous particles depends on the temperature and moisture content [28, 29].

The formation of microparticles takes place when the sprayed droplet enters in contact with the gas (hotter than 100°C). While the moisture is removed, the shell particle is formed, creating structures with different shapes. The vapor produced inside particles expands and collapses the particle structure; therefore, such vapor is primarily responsible for the powder development [30, 31].

When the initial solution is a homogeneous product, it is transformed into two layers; one of them is dense, and the other reveals a porous surface. However, the development of particle microstructure is complex due to the interaction of drying operation conditions and physicochemical liquid-fed composition [31, 32]. When using maltodextrin, changes in the morphological structure of particles are related to the moisture content of material and drying air temperatures. At low temperatures, particles with rough surface were observed; at high temperatures, a smooth surface (but with a greater number of particles broken or fragmented) was developed [31].

4.1. Morphology and microstructure

The particle morphology is described in terms of particle size, shape, surface properties, and internal structure [33]. It is determined by the physical and chemical properties of the particle shell, which at the same time depend on the composition and concentration of power, viscosity, and drying air conditions [34].

Optical microscopy, confocal laser scanning microscopy, and electron microscopy scanning are different techniques used to characterize powder (primary and secondary or agglomerates). Additionally, through digital image analysis, quantitative information is obtained [30, 35]. The scanning electron microscopy (SEM) is one of the most appropriate techniques for the characterization of the morphology and microstructure of food. It is based on the scanning of the sample surface, and it produces results in the form of digital images obtained directly from the microscope. The images can be translated into numeric data for subsequent statistical analysis, and its quantification at any resolution scale may indicate structural changes due to processing [36].

In this sense, digital image analysis is a useful tool that allows using an image to quantitatively describe the different morphometric, colorimetric, and statistical characteristics. Quantification and classification of images via dimensional descriptors such as shape, texture, diameter, and area allow the identification of structural changes at any resolution level. Digital image analysis can be used to evaluate, compare, and characterize microencapsulation under different processing conditions [37, 38].

5. Case study

A fruit of special interest is the xoconostle (*Opuntia matudae*). It is an acid fruit containing a significant amount of polysaccharides, soluble fiber, simple phenols, betalains, and ascorbic acid (which confers antioxidant capacity) [39], among other compounds such as proteins and ashes. In order to improve the sustainable development in rural agricultural areas, the processing of xonocostle could be done by the spray drying of the juice, and the resulting product could be used as a food additive. However, due to its high sugar content, the juice has to be added with a drying adjuvant such as maltodextrin in order to improve the drying process.

5.1. *Opuntia xoconostle* fruit

The xoconostle fruit is a pyriform berry with an apical depression or receptacle. It is composed of the epicarp (skin or shell), mesocarp (pulp), and the endocarp where the seeds are tightly bound in a mucilaginous structure [40] (**Figure 1**). In the genus Opuntia, a wide variety of species that produce xoconostle fruit have been reported, including *Opuntia joconostle* and *Opuntia matudae* (*Xoconostle cuaresmeño*). Xoconostle is an acid fruit that has been underused and considered as an agricultural waste product, although it contains a significant amount of fiber, minerals, phenolic compounds, betalains, organic acids, and waxes [39] with an important nutritional contribution, which makes it an attractive product.

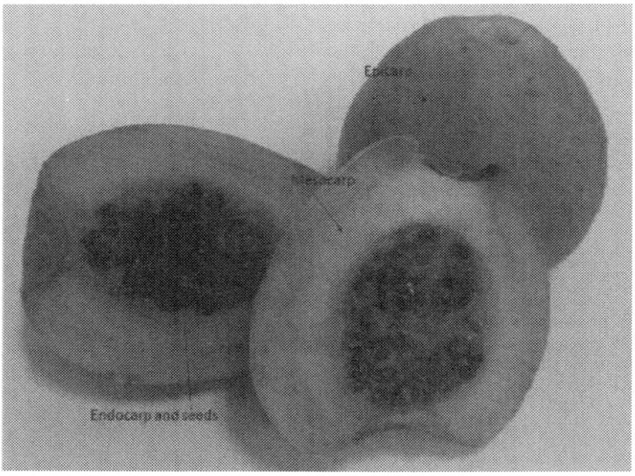

Figure 1. Image of whole and half of *Xoconostle cuaresmeño* fruit. The arrows show the epicarp, mesocarp, endocarp, and seeds.

5.2. Production of xoconostle juice and powder

Xoconostle cuaresmeño (*Opuntia matudae*) of the municipality of San Martín de las Pirámides, Mexico State, Mexico, was collected. The xoconostle batch was put in wooden boxes with kraft paper covering the fruit. Then, it was stored under refrigeration conditions until processing.

Juice extraction was performed by subjecting the fruit clean, after cut into slices, to water bath using the 2:1 ratio (fruit-water) at 70°C for 20 min. Then it was crushed in a pulper machine. The fruit juice was extracted including compounds of the mucilage and seed. Maltodextrin 20DE was added in the ratio 1:1 w/w (with respect to xoconostle total solids). Maltodextrin was used as adjuvant for spray drying juice. After obtaining the blend, an aliquot was separated for analysis, and the rest was spray dried.

A two-fluid nozzle laboratory co-current spray dryer (Mobile Minor 2000, GEA Niro, Denmark) was used, equipped with a peristaltic pump (Model 520, Watson Marlow, USA) for liquid flow control. The spray drying was conducted under 110/60, 155/70, and 200°C/80°C, as inlet/outlet drying air temperature.

5.3. Characterization of xoconostle juice and powder

5.3.1. Degrees Brix (°Bx)

The degrees Brix refers to the amount of dissolved solids extracted from fruit or sugary liquids. The method is based on the change of direction of the light because of the separation of two media in which the velocity of propagation is different. Degrees Brix reading was carried out with a refractometer (Master T, Atago, Japan). Measurements were done in triplicate.

5.3.2. Powder moisture content (MC)

One gram of powder sample was poured into an aluminum tray (constant weight) and was placed in a vacuum oven at 60°C for 24 h [41]. Measurements were done in triplicate.

5.3.3. Water activity (Aw)

One gram of xoconostle powder sample was put in a cell and then subjected to scanner computer activity using a water activity analyzer (AquaLAB 4TE, Decagon Devices, USA), at a constant temperature of 24 ± 1°C. Measurements were done in triplicate.

5.3.4. Glass transition temperature (T_g)

Fifteen milligrams of xoconostle juice powder were poured into an aluminum capsule to be analyzed using a calorimeter (DSC Diamond, Perkin Elmer, USA). The temperature interval of the analysis was from −40 to 120°C, at a heating speed of 5°C/min. The resulting thermograms were analyzed to determine the glass transition temperature (T_g) [42].

5.3.5. Scanning electron microscopy (SEM)

To assess the morphology of microparticles, powders were fixed on carbon tape placed on a specimen slide, and the remainders were removed. Samples were observed by means of a double-beamed scanning electron microscope (Dual Beam Nova 200 Nanolab, FEI) operated at 1.00 kV (for sensible materials), a total amplification of 1000×, 3000×, and 5000× [30]. Measurements were done in triplicate.

5.3.6. Particle size and shape by digital image analysis

To assess the distribution and particle size, powder samples obtained after spray drying were analyzed using software ExpertShape, where the acquisition of images was carried out manually using the light microscope (CILAS 1090-ExpertShape-NT 2107380, France), which has a video camera with a peak bandwidth of 23.2 MB/s. A microscope suited with a 10×

objective lens was used. Illumination was provided by a light-field source. A total of 1300 particles were analyzed for each experiment.

For particle measuring, software ExpertShape acquires the image of a particle defined as a group of contiguous nonzero pixels. Particles can be characterized by a relationship of measures depending on their attributes such as location of the particle, area, and form factor. The coordinates are expressed with reference to the origin (0, 0), located at the upper left corner of the image. When extracting a particle in the image, the outline is defined by the projection of the sensor cutting. Contour is defined by scanning the sequence of adjacent pixels for each particle. The outline is drawn using the chain of "Freeman" code (gray-level image, the thresholder image, the shape without holes and smooth contour, and extracted contour with Freeman chain-code algorithm). Some morphological parameters that could be evaluated are area (A); perimeter (P, as the length of the contour of the particle); *mean Feret diameter* (Calipter diameter) defined as the distance between two tangents on opposite sides of the particle, parallel to some fixed directions, touching opposite sides of particle; *maximum Feret diameter*, as the length of particle; *minimum Feret diameter* is the width of particle; *equivalent circular diameter* is calculated as the circumference with the same area as the projected particle; *roundness* is defined as the ratio of the area of a circle (the most compact shape) that has the same perimeter, and the closer the shape of the particle is to a disk, the closer the roundness is to 1; *equivalent ellipse ratio*, as the ratio of the major axis to its minor axis (elliptical shapes exhibit ratio >1) [43].

5.3.7. Total phenols by Folin-Ciocalteu method

Total phenols of xoconostle juice and powder were obtained by the method of Folin-Ciocalteu to determine the thermal damage caused by drying air temperature as an operation condition of spray drying.

For the extraction of total phenols, 1 g of powder was mixed with 5 mL of 80% methanol and shaken for 30 min at 200 rpm at room temperature. The sample was decanted using filter paper of 110 mm (qualitative circles Whatman 3). The supernatant was separated and placed into an amber glass bottle. The sediment was reconstituted, using 8 mL of 80% ethanol, and the procedure was repeated twice. Thereby, three extractions were joined to be analyzed. For the xoconostle juice, a sample of 4 mL was used, following the previous methodology. For the quantification stage, 0.75 μL of Folin-Ciocalteu reagent (1:10) was added to 100 μL of the extract, and the mixture was left to stand for 5 min in the dark. Subsequently, 0.75 mL of $NaHCO_3$ 60 g/L solution was added to neutralize the reaction. The solution was left to stand for 90 min, and the absorbance was determined at 725 nm. The results were reported as mg gallic acid equivalents/100 g (db) [44]. Measurements were done in triplicate.

5.3.8. Extraction and quantification of betalains

One gram of powder sample was mixed with 5 mL of a solution: methanol ((80:20 v/v), stirring (at maximum speed) using a magnetic stirrer for 30 min and 5°C. The solution was filtered using a membrane pore with a size of 110 mm. The obtained extract was analyzed, expressing the result as mg of betacyanin/100 g of sample (db), since it has been reported that betacyanins are the main components of betalains [45, 46].

5.3.9. Antioxidant activity

ABTS (7 mM) radical cation (ABTS•⁺) solution was prepared by reacting ABTS (2, 2'-azino-bis(3-ethylbenzothiazoline-6-sulphonic acid)) with 2.45 mM potassium persulfate and allowing the mixture to stand in the dark at room temperature for 12–16 h. The ABTS•⁺ radical was diluted with ethanol to give an absorbance of about 0.700±0.02 at 734 nm. In order to measure the antioxidant capacity, 10 μL of the extract was mixed with 990 μL of radical solution. The absorbance was monitored at 734 nm for 7 m. All experiments were carried out in triplicate. The results were expressed in terms of mM Trolox equivalent/g sample (mM TE/g, db) [44, 47].

5.4. Results and discussion

Spray drying of xoconostle juice was conducted with low content of maltodextrin as carrier substances, improving the drying characteristics, avoiding technical problems such as the presence of sticky powder and the accumulation of wet material on the walls of the drying chamber and the mechanical cyclone. Although xoconostle juice contains a very low solid concentration (3°Brix), this value was the reference to choose the quantity of carrier agents to be mixed with the juice. The relation of juice solids-maltodextrin (1:1) was enough to reach yield levels from 42 to 79%, the best drying condition being at 110°C/60°C, to obtain the highest powder recuperation (**Figure 2**). This fact could be affected by the powder density and mean particle size, by keeping constant the concentration of the feed flow and atomizer pressure (0.6 Bars). Low drying air temperature produces compact and small particles, as reported by Alamilla-Beltrán et al. [31].

Figure 2. Images of the xoconostle juice and the xoconostle juice powder obtained by spray drying at 110°C/60°C as inlet/outlet drying air temperatures.

Moisture content, water activity, and glass transition temperature (T_g) were evaluated for all conditions, obtaining values of 4–4.98% (db), 0.19–0.23, and 34.37–38°C, respectively

(**Table 1**). The results showed that powdered xoconostle juice was highly stable due to reduced values of moisture and water activity, improving storage stability with the incorporation of a carrier of high molecular weight such as maltodextrin by increasing the glass transition temperature, and considering that fructose has lower values of T_g (16–17.6°C).

Parameters	Drying air temperature (inlet/outlet) (°C)		
	110/60	155/70	200/80
Moisture content (% db)	4.99	4.07	4.88
Water activity (Aw)	0.19	0.22	0.23
Glass transition temperature (°C)	34.70	34.75	38.45
Feret diameter mean (μm)	14.76	16.11	16.12
Feret diameter max (μm)	12.12	11.12	13.27
Feret diameter min (μm)	16.87	17.76	18.40
Equivalent elliptical ratio	1.94	1.80	1.85
Form factor (roundness)	0.8055	0.8571	0.8453
Phenolic content (mgGAE/100 g)	1250.10	1365.60	1359.10
Betalain content (mg/100 g)	2.65	2.14	2.42
Antioxidant capacity (mM TE/g)	65.08	64.39	67.26

Table 1. Mean values of different parameters evaluated for xoconostle juice powder obtained by spray drying process at three different inlet/outlet drying air temperatures.

This variety of fruit contains compounds such as oxalates, organic acids, vitamins, and wax. In this study, the effect of these has not been analyzed; however, it could exist as an important interaction between wax and maltodextrin, which may improve the protective effect upon powdered xoconostle juice. Regarding the drying temperature, the highest condition improves the reduction on powder moisture content, although this difference is minimal. No evident difference was found in the water activity values, which means that drying temperatures do not affect this parameter, getting low values in all cases. At all combined values of water activity and moisture content obtained in this work, biochemical reaction and enzymatic activity were inhibited, and bacterial growth could be reduced.

During spray drying, mass and heat transfer mechanisms act inducing the evaporation of water; meanwhile, solids and moisture content diffuse from the inside of droplets to the outside, developing solid particles. This migration of components was induced by the air-drying temperature and diffusion factors, to create powders with microstructures that influence the final quality of powders.

Powder microstructure could be analyzed by using diverse methods. In this work, scanning electron microscopy (SEM), optical microscopy, and digital image analysis provided information about morphometric parameters such as size, perimeter, area, shape, and structure of microparticle (**Table 1**). Under all conditions, the powdered xoconostle juice (analyzed with SEM) revealed spherical individual particles (primary particles) forming weak agglomerates (secondary particles). The shape of primary particles tends to be spherical with slightly deep depressions of surface; meanwhile, agglomerates form an elliptical shape, without a structured order. Some particles exhibit an external surface with highly porous aspect. This fact could be caused by a selective migration of solutes dissolved in the xoconostle juice. Solutes with low molecular weight could diffuse quickly to the surface forming the porous shell, which means that sugars may be structured and form the external surface. These kinds of particles are not desirable due to the high hygroscopicity of sugars. For all drying temperatures, particle microstructure was similar, with differences assessed by morphometric parameters.

Morphometric parameters evaluated by optical microscopy and digital image analysis described differences in particles for all experimental conditions. In this case, knowledge on morphometric parameters is useful to improve the particle description, so identifying any change induced by the processing conditions (drying air temperature, atomizing pressure, and concentration of feed flow) is useful to improve the spray drying process. This means that any change in the processing conditions will be translated into physicochemical properties of powders. In this work, the morphological parameters gave information about dimensions and particle shape.

The highest particles measured by mean Feret diameter (16.12 μm), maximum Feret diameter (18.4 μm), minimum Feret diameter (13.3 μm), and area (209 μm^2) were obtained at 200°C/80°C. These results could explain the lower moisture content and water activity values obtained at 200°C/80°C, and due to particle expansion, the moisture diffusion could be facilitated through thin wall particles. The opposite was observed in products obtained at reduced drying air temperatures.

Although primary particles observed by SEM seem to be spherical and the agglomerates showed an elliptical form, form factor (or roundness) and equivalent ellipse ratio were calculated as a descriptor of bulk powder. Values of roundness were close to 0.85 for all conditions, being lightly higher for particles obtained at 200°C/80°C, so it could be related to particle expansion and less formed agglomerates (1.84 as equivalent elliptical ration). In the case of 110°C/60°C, this parameter was 1.94.

For all drying air temperatures, total phenolics (1250–1366 mg GAE/100 g), betalain content (2.14–2.65 mg/100 g), and antioxidant activity (64–67 mM TE/g) did not exhibit significant differences. This means that any option could be applied to produce a powdered xoconostle juice containing phenolic compounds and betalains; however, microstructural, physicochemical, and flow properties have to be considered when a powder with functional properties is designed.

6. Conclusions

The spray drying process has been widely used as a useful tool to generate functional products in the form of powders. The processing of xoconostle juice by spray drying allows proposing an option to strengthen the sustainable use of this product as a natural additive in the food Industry, reinforcing the valorization and reutilization of valuable components found in the agro-industrial waste. By keeping constant the drying conditions (such as feed concentration, atomization pressure, flow arrangement, and type of atomizer) and changing the drying air temperature, the powdered xoconostle juice was obtained.

For all cases, the drying air temperature had an evident positive effect upon physical stability given by the low values of water activity and moisture content, and the increase in T_g by the addition of maltodextrin. The biological functionality (evaluated by total phenolics, betalains, and antioxidant activity) was also kept in similar values for all drying conditions. The most evident changes were observed at microscopic scale. For designing similar products, it is necessary to consider the interactions between food properties (microstructural, functional, physicochemical, and flow) to take the best decision.

Acknowledgements

The authors acknowledge SIP-IPN (Projects 20160284, 20161775, and 20161749); CONACyT-Mexico for financial support (Projects 216044 and 272150-MOR2015); and COFAA-IPN. Jaime Jiménez-Guzmán thanks CONACYT-Mexico and IPN for study grant.

Author details

Jaime Jiménez-Guzmán[1], Diana E. Leyva-Daniel[1], Brenda H. Camacho-Díaz[2], Antonio R. Jimenéz-Aparicio[2] and Liliana Alamilla-Beltrán[1*]

*Address all correspondence to: liliana.alamilla@gmail.com, lalamill@ipn.mx

1 National School of Biological Sciences-National Polytechnic Institute, Department of Biochemical Engineering, CP, México

2 Biotic Products Development Center, National Polytechnic Institute, Yautepec, Morelos, México

References

[1] Sougnez, M. The development of the spray drying (L'évolutione du séchage par atomisation). Chim. Mag. 1983, 1, 1–4.

[2] Allen, R, Bakker, H. Spray dryer control based on-line particle size analysis. Chem. Eng. Res. Des. 1994, 72: 251–254. DOI: 10.1017/CBO9781107415324.004.

[3] Oakley, DE. Scale-up of spray dryers with the aid of computational fluid dynamics. Dry. Technol. 1994, 12: 217–233. DOI: 10.1080/07373939408959954.

[4] Goula, AM, Adamopoulos, KG. Effect of maltodextrin addition during spray drying of tomato pulp in dehumidified air: I. Drying kinetics and product recovery. Dry. Technol. 2008, 26: 714–725. DOI: 10.1080/07373930802046369.

[5] Adhikari, B, Howes, T, Bhandari, B, Troung, V. Effect of addition of maltodextrin on drying kinetics and stickiness of sugar and acid-rich foods during convective drying: experiments and modelling. J. Food Eng. 2004, 62: 53–68. DOI: 10.1016/S0260-8774(03)00171-7.

[6] Langrish, TAG, Kota, K. A comparison of collision kernels for sprays from one and two-nozzle atomisation systems. Chem. Eng. J. 2007, 126: 131–138. DOI: 10.1016/j.cej.2006.07.015.

[7] Muzaffar, K, Kumar, P. Moisture sorption isotherms and storage study of spray dried tamarind pulp powder. Powder Technol. 2016, 291: 322–327. DOI: 10.1016/j.powtec.2015.12.046.

[8] Samborska K, Witrowa-Rajchert, D, Goncalves, A. Spray-drying of alpha-amylase – the effect of process variables on the enzyme inactivation. Dry. Technol. 2005, 23: 941–953. DOI: 10.1081/DRT-200054243

[9] Tan, SP, Kha, TC, Parks, SE, Stathopoulos, CE, Roach, PD. Effects of the spray-drying temperatures on the physiochemical properties of an encapsulated bitter melon aqueous extract powder. Powder Technol. 2015, 281: 65–75. DOI: 10.1016/j.powtec.2015.04.074.

[10] Ezhilarasi, PN, Indrani, D, Jena, BS, Anandharamakrishnan, C. Microencapsulation of Garcinia fruit extract by spray drying and its effect on bread quality. J. Sci. Food Agric. 2014, 94: 1116–1123. DOI: 10.1002/jsfa.6378.

[11] Adhikari, B, Howes, T, Bhandari, BR, Truong, V. Experimental studies and kinetics of single drop drying and their relevance in drying of sugar-rich foods: a review. Int. J. Food Prop. 2000, 3: 323–351. DOI: 10.1080/10942910009524639.

[12] Risch, S. Encapsulation: overview of uses and techniques. ACS Symp. Ser. 1995, 7: 2–7. DOI: doi:10.1021/bk-1995-0590.ch001.

[13] Okuyama, K, Wuled Lenggoro, I. Preparation of nanoparticles via spray route. Chem. Eng. Sci. 2003, 58: 537–547. DOI: 10.1016/S0009-2509(02)00578-X.

[14] Jafari, SM, He, Y, Bhandari, B. Encapsulation of nanoparticles of d-limonene by spray drying: role of emulsifiers and emulsifying techniques. Dry. Technol. 2007, 25: 1069–1079. DOI: 10.1080/07373930701396758.

[15] Lefebvre, AH. Atomization [Internet]. 2011. Available from: http://www.thermopedia.com/content/573/ [Accessed: 2016-01-15]

[16] Soottitantawat, A, Bigeard, F, Yoshii, H, Furuta, T, Ohkawara, M, Linko, P. Influence of emulsion and powder size on the stability of encapsulated d-limonene by spray drying. Innov. Food Sci. Emerg. Technol. 2005, 6: 107–114. DOI: 10.1016/j.ifset.2004.09.003.

[17] Kim, EH-J, Chen, XD, Pearce, D. Surface composition of industrial spray-dried milk powders. 2. Effects of spray drying conditions on the surface composition. J. Food Eng. 2009, 94: 169–181. DOI: 10.1016/j.jfoodeng.2008.10.020.

[18] Peighambardoust, SH, Golshan Tafti, A, Hesari, J. Application of spray drying for preservation of lactic acid starter cultures: a review. Trends Food Sci. Technol. 2011, 22: 215–224. DOI: 10.1016/j.tifs.2011.01.009.

[19] Mezhericher, M, Levy, A, Borde, I. Heat and mass transfer of single droplet/wet particle drying. Chem. Eng. Sci. 2008, 63: 12–23. DOI: 10.1016/j.ces.2007.08.052.

[20] Bilancetti, L, Poncelet, D, Loisel, C, Mazzitelli, S, Nastruzzi, C. A statistical approach to optimize the spray drying of starch particles: application to dry powder coating. AAPS PharmSciTech. 2010, 11: 1257–1267. DOI: 10.1208/s12249-010-9492-y.

[21] Mezhericher, M, Levy, A, Borde, I. Spray drying modelling based on advanced droplet drying kinetics. Chem. Eng. Process. Process Intensif. 2010, 49: 1205–1213. DOI: 10.1016/j.cep.2010.09.002.

[22] Handscomb, CS, Kraft, MÃ. Simulating the structural evolution of droplets following shell formation. Chem. Eng. Sci. 2010, 65: 713–725. DOI: 10.1016/j.ces.2009.09.025.

[23] Bayu, A, Nandiyanto, D, Okuyama, K. Progress in developing spray-drying methods for the production of controlled morphology particles: from the nanometer to submicrometer size ranges. Adv. Powder Technol. 2011, 22: 1–19. DOI: 10.1016/j.apt.2010.09.011

[24] Boonyai, P, Bhandari, B, Howes, T. Stickiness measurement techniques for food powders: a review. Powder Technol. 2004, 145: 34–46. DOI: 10.1016/j.powtec.2004.04.039.

[25] Mani, S, Jaya, S, Das, H. Sticky issues on spray drying of fruit juices. In:North-Central International meeting sponsored by ASAE and CSAE (2002 ASAE/CSAE); 27–28 September 2002; Saskatoon, Saskatchewan, Canada. 2002. p. 1–18

[26] Roos, Y. Importance of glass transition and water activity to spray drying and stability of dairy powders. Lait 2002, 82: 475–484. DOI: 10.1051/lait:2002025.

[27] Turchiuli, C. Fluidization in food powder production. In: Bhandari, B., Bansal, N., Zhang, M., Schuck, P, editors, Handbook of Food Powders: Processes and Properties.

1st ed. Woodhead Publishing Limited, 2013. p. 178–199. DOI: 10.1533/9780857098672.1.178

[28] Woo, MW, Daud, WRW, Tasirin, SM, Talib, MZM. Effect of wall surface properties at different drying kinetics on the deposition problem in spray drying. Dry. Technol. 2008, 26: 15–26. DOI: 10.1080/07373930701781033.

[29] Adhikari, B, Howes, T, Lecomte, D, Bhandari, BR. A glass transition temperature approach for the prediction of the surface stickiness of a drying droplet during spray drying. Powder Technol. 2005, 149: 168–179. DOI: 10.1016/j.powtec.2004.11.007.

[30] Porras-Saavedra, J, Palacios-Gonzalez, E, Lartundo-Rojas, L, Garibay-Febles, V, Yañez-Fernández, J, Hernández-Sánchez, H, Gutiérrez-López, G, Alamilla-Beltrán, L. Microstructural properties and distribution of components in microparticles obtained by spray-drying. J. Food Eng. 2015, 152: 105–112. DOI: 10.1016/j.jfoodeng.2014.11.014.

[31] Alamilla-Beltrán, L, Chanona-Pérez, JJ, Jiménez-Aparicio, AR, Gutiérrez-López, GF. Description of morphological changes of particles along spray drying. J. Food Eng. 2005, 67: 179–184. DOI: 10.1016/j.jfoodeng.2004.05.063.

[32] Walton, DE, Member, CJM. Spray dried products-characterization of particle morphology. Trans. IChemE 1999, 77: 21–38.

[33] Vehring, R. Pharmaceutical particle engineering via spray drying. Pharm. Res. 2008, 25: 999–1022. DOI: 10.1007/s11095-007-9475-1.

[34] Paramita, V, Futura, T, Yoshii, H. Microencapsulation efficacy of d-limonene by spray drying using various combinations of wall materials and emulsifiers. Food Sci. Technol. Int. Tokyo 2010, 16: 365–372. DOI: 10.3136/fstr.16.365.

[35] Schoonman, A, Mayor, G, Dillmann, ML, Bisperink, C, Ubbink, J. The microstructure of foamed maltodextrin/sodium caseinate powders: a comparative study by microscopy and physical techniques. Food Res. Int. 2001, 34: 913–929. DOI: 10.1016/S0963-9969(01)00116-8.

[36] Gonzalez, AL, Noguez, C. Influence of morphology on the optical properties of metal nanoparticles. J. Comput. Theor. Nanosci. 2006, 4: 231–238.

[37] Quintanilla-Carvajal, MX, Meraz-Torres, LS, Alamilla-Beltrán, L, Chanona-Pérez, JJ, Terres-Rojas, E, Jiménez-Aparicio, AR. Morphometric characterization of spray-dried microcapsules before and after A-tocopherol extraction (Caracterizac on Morfo Etrica De Microcapsulas Secadas Por Aspers on Antes Y Desp Es De La Extracc on De A-Tocoferol). Rev. Mex. Ing. Química 2009, 8: 301–312.

[38] Meraz-Torres, LS, Quintanilla-Carvajal, MX, Hernández-Sánchez, H, Téllez-Medina, DI, Alamilla-Beltrán, L, Gutiérrez-López, GF. Assessment of the kinetics of contact angle during the wetting of maltodextrin agglomerates. Rev. Mex. Ing. Química 2011, 10: 273–279.

[39] Morales, P, Ramírez-Moreno, E, Sanchez-Mata, M de C, Carvalho, AM, Ferreira, ICFR. Nutritional and antioxidant properties of pulp and seeds of two xoconostle cultivars (Opuntia joconostle F.A.C. Weber ex Diguet and *Opuntia matudae* Scheinvar) of high consumption in Mexico. Food Res. Int. 2012, 46: 279–285. DOI: 10.1016/j.foodres.2011.12.031.

[40] Osorio-Esquivel, O, Alicia-Ortiz-Moreno, Álvarez, VB, Dorantes-Álvarez, L, Giusti, MM. Phenolics, betacyanins and antioxidant activity in *Opuntia joconostle* fruits. Food Res. Int. 2011, 44: 2160–2168. DOI: 10.1016/j.foodres.2011.02.011.

[41] AOAC. Official methods of analysis Method 930.15. 18th ed. Asociation of official analytical chemists; Arlington, VA, USA. 2005

[42] Ozmen, L, Langrish, TAG. Comparison of glass transition temperature and sticky point temperature for skim milkpowder. Dry Technol. 2002, 20: 1177–1192. DOI: 10.1081/DRT-120004046.

[43] CILAS. Expertshape Software: Instruction Manual. Orleans Cedex, France. 2011

[44] Jaramillo-Flores, ME, González-Cruz, L, Cornejo-Mazón, M, Dorantes-Alvarez, L, Gutiérrez-López, GF, Hernández-Sánchez, H. Effect of thermal treatment on the antioxidant activity and content of carotenoids and phenolic compounds of cactus pear cladodes (*Opuntia ficus-indica*). Food Sci. Technol. Int. 2003, 9: 271–278. DOI: 10.1177/108201303036093.

[45] Stintzing, FC, Herbach, KM, Mosshammer, MR, Carle, R, Yi, W, Sellappan, S, Akoh, CC, Bunch, R, Felker, P. Color, betalain pattern, and antioxidant properties of cactus pear (*Opuntia* spp.) clones. J. Agric. Food Chem. 2005, 53: 442–451. DOI: 10.1021/jf048751y.

[46] Sánchez-González, N. Extraction and characterization of major pigments Opuntia joconoste c.v. (Extracción y caracterización de los principales pigmentos del Opuntia joconoste c.v.) (xoconostle). Master thesis. 2006, 1–107.

[47] Re, R, Pellegrini, N, Proteggente, A, Pannala, A, Yang, M, Rice-Evans, C. Antioxidant activity applying an improved ABTS radical cation decolorization assay. Free Radic. Biol. Med. 1999, 26: 1231–1237. DOI: 10.1016/S0891-5849(98)00315-3.